U.S. Marine Corps Combat Water Survival

Plus

Soldier's Handbook For Individual Operations Survival In Cold Weather Areas

Department of Defense

Pentagon
Publishing

MCRP 3-02C

Marine Combat Water Survival

U.S. Marine Corps

PCN 144 000069 00

To Our Readers

Changes: Readers of this publication are encouraged to submit suggestions and changes that will improve it. Recommendations may be sent directly to Commanding General, Marine Corps Combat Development Command, Doctrine Division (C 42), 3300 Russell Road, Suite 318A, Quantico, VA 22134-5021 or by fax to 703-784-2917 (DSN 278-2917) or by E-mail to **morgann@mccdc.usmc.mil**. Recommendations should include the following information:

- Location of change
  Publication number and title
  Current page number
  Paragraph number (if applicable)
  Line number
  Figure or table number (if applicable)
- Nature of change
  Add, delete
  Proposed new text, preferably double-spaced and typewritten
- Justification and/or source of change

Additional copies: A printed copy of this publication may be obtained from Marine Corps Logistics Base, Albany, GA 31704-5001, by following the instructions in MCBul 5600, *Marine Corps Doctrinal Publications Status.* An electronic copy may be obtained from the Doctrine Division, MCCDC, world wide web home page which is found at the following universal reference locator: **http://www.doctrine.usmc.mil**.

Unless otherwise stated, whenever the masculine gender is used, both men and women are included.

DEPARTMENT OF THE NAVY
Headquarters United States Marine Corps
Washington, D.C. 20380-1775

6 January 2003

FOREWORD

Marine Corps Reference Publication (MCRP) 3-02C, *Marine Combat Water Survival*, provides Marine Corps combat water survival techniques, procedures, and training standards. This publication also teaches Marines to cross water obstacles and perform water rescues correctly and safely.

This publication is the foundation for teaching Marines correct water survival techniques and procedures that are used throughout the Marine combat water survival program (MCWSP). Once an individual or a unit has completed the MCWSP, this publication can be used as a refresher course before water operations.

MCRP 3-02C supersedes Fleet Marine Force Reference Publication (FMFRP) 0-13, *Marine Combat Water Survival*, dated 16 September 1991.

Reviewed and approved this date.

BY DIRECTION OF THE COMMANDANT OF THE MARINE CORPS

EDWARD HANLON, JR.
Lieutenant General, U.S. Marine Corps
Commanding General
Marine Corps Combat Development Command

Publication Control Number: 144 000069 00

MCRP 3-02C Marine Combat Water Survival

Table of Contents

Chapter 1. Survival at Sea

Chapter 2. Water Rescues

Chapter 3. Treatment of Casualties and Avoidance of Dangerous Marine Life

Chapter 4. Negotiating Water Obstacles

Chapter 5. Fording Waterways

Appendix. Knot Tying

Throughout history, water has posed special challenges to Marines and Sailors during times of both peace and war. Therefore, the inherent dangers associated with waterborne operations demand that Marines and Sailors receive proper water survival training. Combat units that are confident in their ability to work in and around water can use the water to their advantage in combat, and history is filled with examples of how the proper preparation or training for survival in water reduced or averted disaster.

USS *Indianapolis*

On Sunday, 29 July 29 1945, the heavy cruiser USS *Indianapolis* (the *Indy*) was en route to the Philippine Sea. Shortly before midnight (about 39 hours out of port), the *Indy*, running blacked out and unescorted, was rocked by two explosions on her starboard side. With communications smashed, the ship could not signal its distress and sank within 15 minutes.

Three life rafts and a floater net supported a few survivors, but the rest of the survivors drifted about, held up by rubber life belts or Mae Wests. About 60 seamen died the first night. Survivors assumed the ship would be reported overdue in Leyte, and that they would be rescued within 2 days. By Monday evening, however, panic began to set in as some life jackets lost their buoyancy from the long immersion. Some men even fought over life jackets, which resulted in at least 25 deaths. No one dared sleep for fear of losing his jacket.

Throughout the next several days, in-transit aircraft flew nearby without spotting the desperate seamen. As best they could, the men kept together, some tying long ropes to each other, floating like corks on a net.

Not until late Thursday morning, 3 1/2 days after the ship sank, were the men discovered. Luckily, a plane on a routine flight over the area spotted the survivors. When ships picked them up that night, the survivors learned they had never been reported overdue. Every one of the 1,196-man crew was a casualty; 880 crewmembers were listed as dead or missing. Although many lives were lost, the innovative and expedient use of flotation devices and float techniques employed by the survivors helped save hundreds of lives.

USS *America*

On Thanksgiving eve, 23 November 1995, the USS *America* made its way through the Arabian Sea. Twenty-year-old, Marine Lance Corporal (LCpl) Zachary Mayo was unable to sleep and, wanting some fresh air, made his way onto an open-air platform near the aircraft hangar bay, which was three levels below the sleeping quarters. While he was on the platform, the ship veered suddenly, throwing LCpl Mayo through the platform's protective bars and into the sea, 30 feet below.

Frantic, LCpl Mayo called out in vain to the watchmen on the flight deck, which was 64 feet above him. It soon became clear to him that the USS *America* would keep its course into the Gulf of Oman until his absence was discovered at morning muster. The LCpl took a moment to consider his situation. Since land was at least 100 miles away, swimming was suicide; he would have to stay afloat until a search party found him.

Using the techniques he had learned during combat water survival training, LCpl Mayo made a flotation device out of his coveralls and tried to relax. Meanwhile, business continued as usual aboard the USS *America*. Since LCpl Mayo was on special assignment with the hazardous materials division, his absence went unnoticed until a petty officer asked several Sailors if they had seen their shipmate recently. By the time a roll call had been completed, LCpl Mayo had been adrift at sea for over 24 hours. Although three, fixed-wing Viking aircraft were deployed to search for LCpl Mayo, most people aboard ship feared the worst.

After 34 hours at sea, LCpl Mayo was discovered by fishermen on a Pakistani fishing boat. LCpl Mayo's survival is a testament not only to his incredible physical courage, but also to the soundness of the lifesaving training and techniques he received during combat water survival training.

Chapter 1
Survival at Sea

As a Marine, you face a variety of potential water emergencies whenever you cross expanses of water: ships, watercraft, and amphibious assault vehicles (AAVs) can sink; aircraft can crash into the sea; or you can accidentally fall into the water. However, there are some basic precautionary measures you can take to protect your safety and reduce your chances of becoming a water casualty. Determine the following information as soon as you board any type of vessel. Your knowing the following information may save your life or the life of your fellow shipmates.

- How many life preservers and lifeboats/rafts are on board?
- Where are the life preservers and lifeboats/rafts located?
- What type of unit survival equipment is on board?
- Are individual survival kits issued to each person on board?
- How much food, water, and medicine do the survival kits contain? When was the last time the contents were inspected for proper quantities and shelf life expiration?
- Is there sufficient survival equipment available for the number of personnel?
- How many other personnel are there on board, and where are they located?
- What are the egress procedures for the ship, boat, watercraft, AAV, or aircraft?

Abandoning Ship

When you embark on a Navy ship, you will receive abandoning ship instructions from Navy personnel. If given the order to abandon ship, report to your designated assembly area and put on a life preserver. DO NOT inflate the life preserver until you are clear of the ship. Torn life preservers will not inflate and inflated life preservers can block you, and those behind you, from exiting the ship. A flotation device that has been inflated may also burst if you jump from a significant height. See pages 1-15 through 1-35 for staying afloat with and without a life preserver.

DO NOT remove your clothing, boots, or shoes before abandoning ship. Your trousers and blouse may be the only flotation devices available if your life preserver is faulty or becomes damaged, and your clothes can provide some insulation from the cold water. However, remove your soft cover and place it in a cargo pocket for later use. The soft cover is both lightweight and a good protection against sunburn caused by the sun's rays reflecting off the water.

Jettisoning Equipment

Equipment should be kept properly packed and waterproofed in case you have to abandon ship. If entering the water from a height greater than 30 feet, wearing your equipment (e.g., pack, helmet, gas mask) could cause injury. Upon impact with the water, the helmet will "cup" air inside of it. The chin strap may also create a "hanging effect" as you submerge from the force of the fall. Therefore, you should remove your helmet and gas mask before abandoning ship.

If you are unable to maintain buoyancy due to the amount of equipment secured to your pack and body, then jettisoning some of your equipment may become necessary. Equipment that you should always retain includes canteens of freshwater, first aid kit, soft cover, and survival kit. The survival kit should include first aid items, water purification tablets or drops, fire starting equipment, signalling items (e.g., flashlight, strobe light, chemlights), food procurement items, and shelter items. Other items in a survival kit include sunburn lotion and lip balm, knife, goggles/sunglasses, plastic bag, matches and lighter, mirror. See Marine Corps Reference Publication (MCRP) 3-02F, *Survival*, for a detailed list of survival items and applications.

Abandoning Ship Technique

When abandoning ship, safety considerations must be observed. Use the following technique when abandoning ship without your combat gear:

Place your hands on their opposite shoulders, forming a criss-cross pattern.

Step to the edge of the ship's deck and check the water below for debris or survivors. If the water is clear, look straight ahead and prepare to jump. If the water is not clear, move to another location.

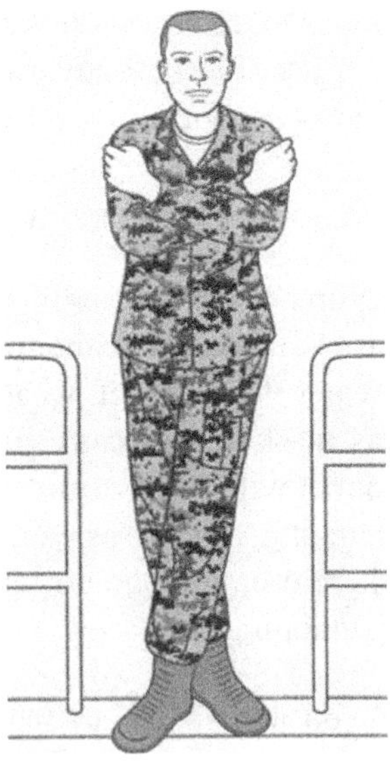

NOTE: DO NOT hold your nose as you abandon ship. If you do hold your nose, the force of impact into the water could jar your arm and hand and cause you to break your nose.

Step off the side of the ship with a smooth, 30-inch stride. DO NOT DIVE OFF THE SHIP. DO NOT LOOK DOWN AT THE WATER. LOOK STRAIGHT AHEAD. Looking down at the water can render you unconscious or cause injuries upon impact.

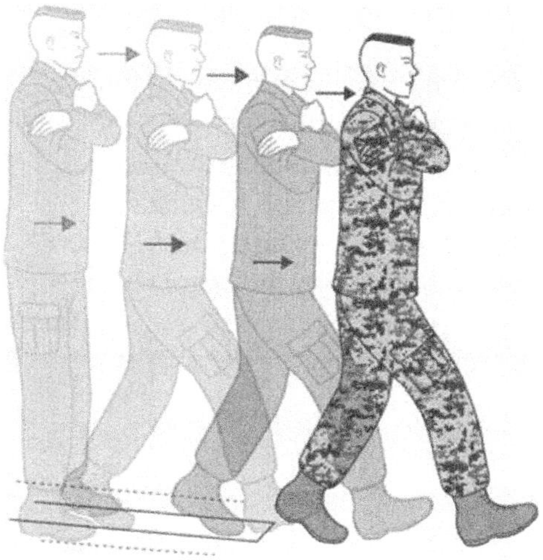

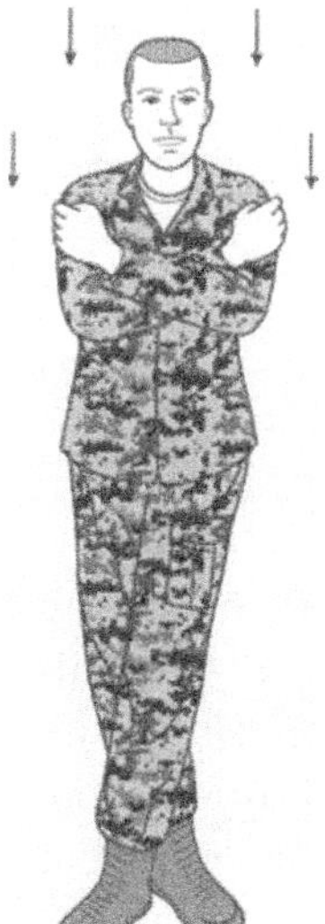

Bring your trailing leg forward during the fall. Cross your trailing leg behind your leading leg.

Keep your head parallel to the water's surface until hitting the water.

You should remain in the abandon ship position until your descent into the water has almost stopped. However, the weight imbalances in your body may cause you to be in a "J" shape under the water

.

Once your downward motion has ceased, your feet may be parallel with the ocean bottom or you may be nearly inverted with your feet over your head. To counteract potential disorientation, you should pause briefly and allow the natural buoyancy of your torso to bring your body to a nearly upright position.

Floating debris can cause hazards. Therefore, you should swim upward, extending one arm (hand is shaped as a fist) upward to feel for obstructions. If you encounter debris, try to push it away or surface in a different location.

Swim away from the ship. DO NOT LOOK BACK AT THE SHIP. Looking back could slow your movement away from the area. Remember, your objective is to leave the area as quickly as possible because—

- Equipment and debris may be falling from or spilling out of the ship.
- Additional casualties can occur if individuals abandoning the ship fall on top of swimmers already in the water.
- Swimmers close to the sinking ship may get pulled underneath the water by the suctioning effect of the ship as it goes under.

Modified Abandoning Ship Technique

When abandoning ship while wearing full combat gear (weapon, helmet, and a properly waterproofed pack), safety considerations must be observed. The modified abandoning ship technique is used while wearing full combat gear and exiting from a height less than 30 feet.

WARNING

When exiting from a height greater than 30 feet, remove your helmet and pack. Fasten the helmet to the pack or place it inside the pack before jettisoning. If jettisoning gear from the ship or aircraft, check the water below for survivors before throwing the gear forward of the intended jump area. Once in the water, retrieve your gear and swim out of the area.

Place your weapon over one shoulder, muzzle down, with the weapon parallel to your side.

Place your arm and hand along the weapon and hold it to your side.

Take your free hand and place it on top of your helmet to prevent neck/spinal injury from the force of the water pulling upward on the helmet as your body enters the water.

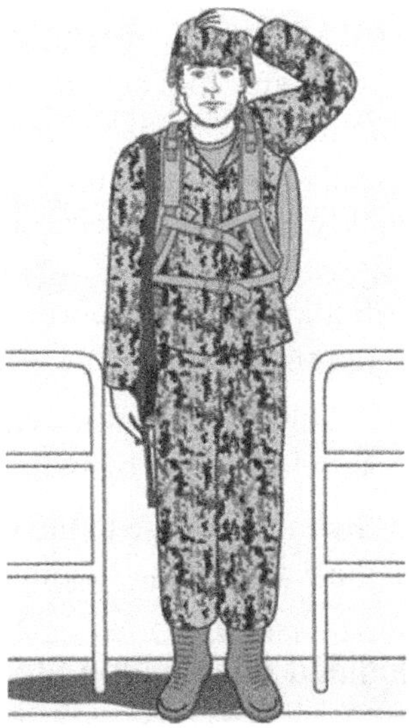

NOTE: An alternate method is to place your weapon over one shoulder, muzzle down, with the weapon parallel to your side. Reach across your body and grasp the sling of your weapon and hold it to your body. Take your free hand and place it on top of your helmet.

Step to the edge of the platform and check the water below for debris and survivors. If the water is clear, look straight ahead and prepare to jump. If the water is not clear, move to another location.

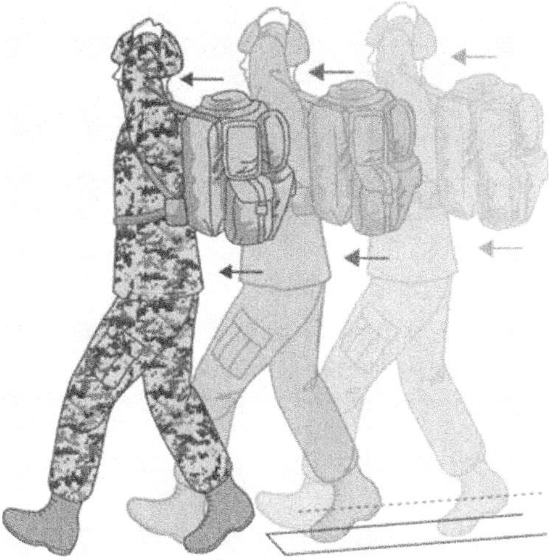

Step off the side of the platform with a smooth, 30-inch stride. DO NOT DIVE OFF THE SHIP, DO NOT LOOK DOWN AT THE WATER, LOOK STRAIGHT AHEAD. Looking down at the water can render you unconscious or cause injury upon impact.

Bring your trailing leg forward during the fall. Cross your trailing leg behind your leading leg.

Keep your head parallel to the water's surface until hitting the water.

You should remain in the modified abandoning ship position until your descent into the water has almost stopped. The buoyancy of a properly waterproofed pack will immediately pull you to the surface. Once you break the water's surface, unsling your weapon and loop the sling over your head. You may have to seesaw the sling so as to ensure the sling passes between your helmet and pack. Once over your head, loop the sling around your neck with the weapon aligned in the center of your body, muzzle down. Lean back on your pack and perform the combat travel stroke to exit the area.

> *Note:* You may also remove the pack (unsnap the quick-release strap on one side of the pack, and pass the strap to your other hand in order to maintain contact with the pack, which is functioning as your flotation device). Remain horizontal in the water with your pack under your chest. Perform a breast stroke kick.

Surface Burning Oil Swim

After you have abandoned ship, rise to the surface using the techniques shown on the page 1-6. However, you must remember that fuel from sinking ships or downed aircraft will float on the surface of the water. Therefore, you must move clear of the floating fuel by swimming away from the ship or aircraft as soon as possible. Either swim upwind (into the wind) of the ship/aircraft or swim against the current. Either method allows you to move away from the fuel and the wind/current will push the fuel past you. To properly execute a surface burning oil swim—

Extend your arms overhead as far as possible.

Wave your arms back and forth vigorously to splash a hole while moving upward.

Splash as long as possible to push burning fuel away from the surfacing area.

Use your arms and hands to sweep away fuel and debris.

Kick your legs in a constant breast stroke kick.

Extend your arms (palms outward) forward on the surface, arms shoulder-width apart.

Pull your hands in and back toward the chest.

Stop your hands in front of your face and rotate them so that your palms face forward (roughly halfway out of the water).

Sweep your arms forward to a full extension at the shoulder width. This splashes debris, oil, or burning liquids aside. To reduce the chance of fatigue, use two short splashes to the front to extend the path.

Repeat the preceding step as necessary while swimming clear of the area.

Surviving With a Pack

If packed properly, your pack will float, and it is your key piece of equipment for staying afloat and overcoming water obstacles. If the pack's contents are properly waterproofed, it can support you (with a combat load) in the water. Buoyed up by a waterproofed pack, you eventually emerge from the water with all your equipment (e.g., boots, helmet, flak jacket, weapon, survival items).

Your pack floats based on a scientific principle known as Archimedes' principle. This principle states that an object submerged in a liquid is buoyed up by a force equal to the weight of the liquid displaced (pushed aside) by the object. If the weight of the displaced liquid is greater than the weight of the object, the object floats. If the weight of the displaced liquid is less than the weight of the object, the object sinks. For example, a machine gun sinks in the water, but it still weighs less in the water than it does on land. Even though a machine gun sinks, it is still buoyed up by a force equal to the weight of the water it displaces. For this reason, you should not try to hold yourself or your equipment any higher out of the water than they would naturally float; doing so wastes both energy and body heat.

Preparing Equipment

Before packing your pack, you must prepare your gear/equipment. Tape or pad all sharp edges and equipment corners. Ideally, your gear/equipment is placed in plastic bags and the plastic bag is then placed inside the standard-issued, rubberized, waterproof bag. Waterproof bags are not completely water tight; henceforth, the added protection of first wrapping the gear/equipment in a plastic bag, then placing the plastic bag inside the waterproof bag. Large plastic bags (e.g., trash bags) work well for bulky equipment (e.g., sleeping bags, field jackets, shelter halves, gas masks). Small plastic bags work best for small items (e.g., shaving gear).

> *NOTE:* If a gas mask must be carried outside the pack, cover it with a waterproof bag.

You will need the following items, which are available through the supply system, to prepare your equipment for packing:

- Waterproof bags:
  Pistol bag, plastic size #1 (8" x 18").
  Rifle bag, plastic size #2 (10" x 56").
  Machine gun bag, plastic size #3 (15" x 56").
  Multipurpose bag, cover, plastic size #4 (20" x 84").

- Panama work vest (life preserver).
- Small plastic bag, self sealing (6" x 6" and 12" x 12").
- Riggers tape, olive drab (2 1/2").
- Willie peter bag, waterproof clothing.

Tying Waterproof/Plastic Bags

Try to remove the excess air from the waterproof/plastic bag before securing the bag's opening. If filled with air, the bag can burst if pressed from the outside. Perform the following steps to tie a waterproof/plastic bag:

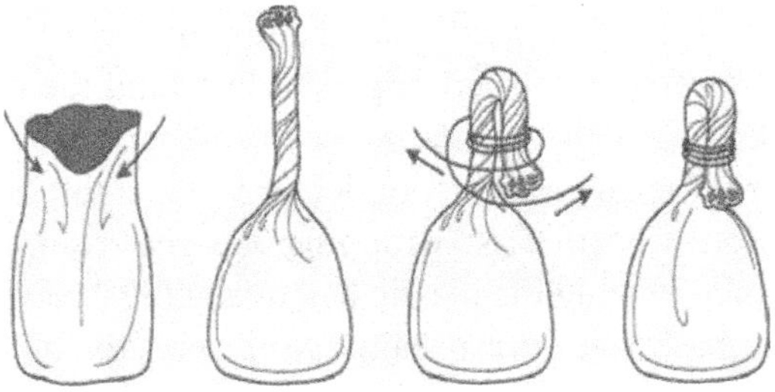

Packing the Pack

Place filled and tied bags inside the pack. Carefully handle sharp items (e.g., tent pegs, poles) to prevent puncturing the bag. Place items in the pack in order of expected use. Close the pack and its compartments. Attach sleeping mats or bags as high as possible on the outside of the pack.

Swimming With the Pack

A useful technique for propelling yourself forward while wearing a pack is the combat travel stroke. This technique allows you to float nearly horizontal and to propel yourself forward with bicycle-style kicks and breast stroke-style arm sweeps. To execute the combat travel stroke—

Body position
Keep the upper part of your body prone to the water, let your legs dangle horizontally, and keep your face up.

Arm action
Extend your hands out in front of your waist. Sweep your arms slightly downward and back 90 degrees to propel your body through the water. Move your hands back to the front of your waist. Repeat.

Leg action
Continuously move your legs in a bicycle-like movement, bringing your knees up high and step out.

Breathing
Keep your face out of the water during the stroke, and breathe freely.

Staying Afloat With a Life Preserver

The best form of flotation is to find any kind of floating object that will keep you and your equipment out of the water or minimize your exposure to the water. Life preservers are the best method, as they allow you to wear your clothes for heat retention and sunburn prevention. Marines use two basic classes of life preservers: inherently buoyant life preservers and inflatable life preservers.

Inherently Buoyant Life Preservers

Inherently buoyant life preservers are either vest-type (worn like a jacket) or yoke-type (worn around the neck). The preserver's outer envelope is either a cotton or water resistant material that encloses a removable fibrous glass or plastic foam filling.

In the Marine Corps, the most common type of inherently buoyant life preserver is the vest-type with collar, known as the kapock preserver. The kapock consists of collar straps, upper front chest straps, leg straps, and waist drawstrings that secure the preserver to you. The leg straps, which are fitted on both sides of the life preserver, ensure that the preserver remains around your chest while you are in the water. A chest strap is attached to the life preserver to facilitate lifting you out of the water. The strap can also be attached to other survivors or to lifeboats to reduce the fatigue that results from holding onto a floating/secured object by hand.

Inflatable Life Preservers

Marine Corps aircraft and AAVs have inflatable life preservers on board. The inflatable life preserver issued to Marines is known as the LPP (life preserver personal). The LPP is capable of both oral inflation and CO_2 cartridge inflation. The LPP consists of buoyancy chambers, CO_2 inflator, and an oral inflation tube. The buoyancy chamber is deck gray in color and is made from a neoprene-coated nylon fabric.

Inflatable life preservers must be stored in a cool, dry place. Heat, moisture, and light cause deterioration of the life preserver material. Do not stow CO_2 cylinders near steam lines or radiators. Heat can increase the pressure inside the cylinders causing them to explode. Avoid sharp edges in stowage. Sharp edges increase wear and tear on the life preservers and may also puncture inflatable buoyancy chambers.

> **CAUTION:** Do not inflate the life preserver until you are clear of the aircraft, ship, or vehicle. Torn life preservers will not inflate and inflated life preservers can block you, and those behind you, from exiting the aircraft, ship, or vehicle.

To don and adjust the LPP—

Remove the LPP from its storage container.

Fasten the belt fasteners in front with the pouch in the rear.

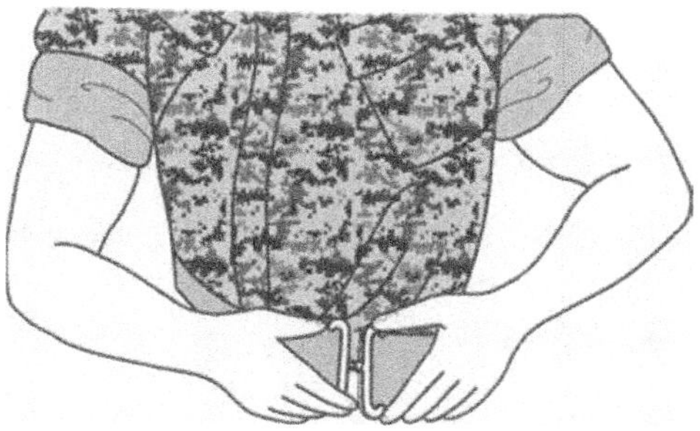

Adjust the belt to fit; secure any excess belt by mating the hook and pile tape.

Rotate the pouch to the front and re-adjust the belt if necessary.

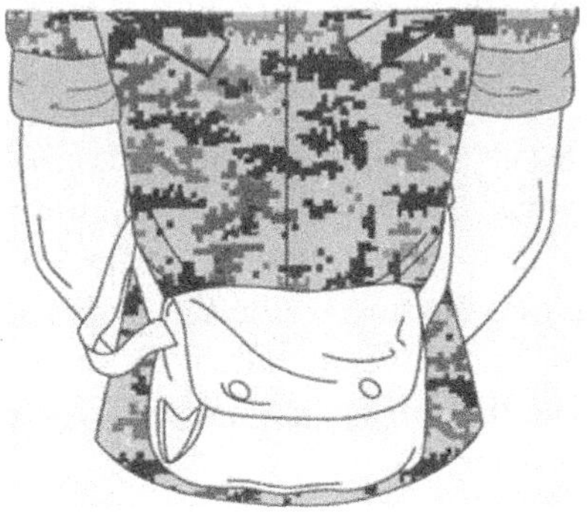

Open the snap fasteners on the pouch and unfold the life preserver.

Place the deflated life preserver over your head.

Place the storage container into the pouch after donning the life preserver.

Lift the lower end of the life preserver out of the pouch.

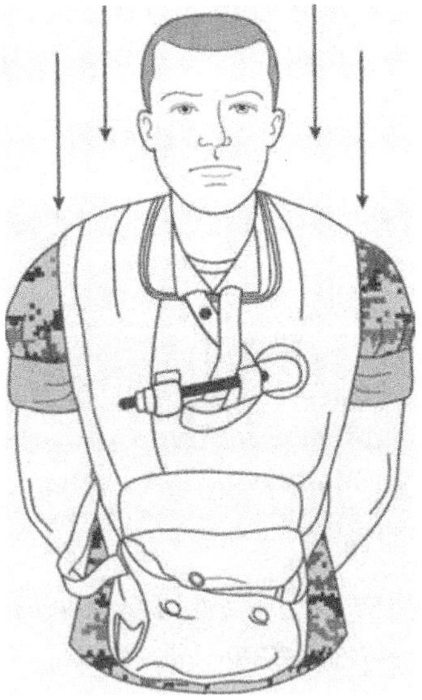

Inflate the life preserver by pulling on the lanyard attached to the CO_2 inflation valve or by blowing on the end of the oral inflation valve.

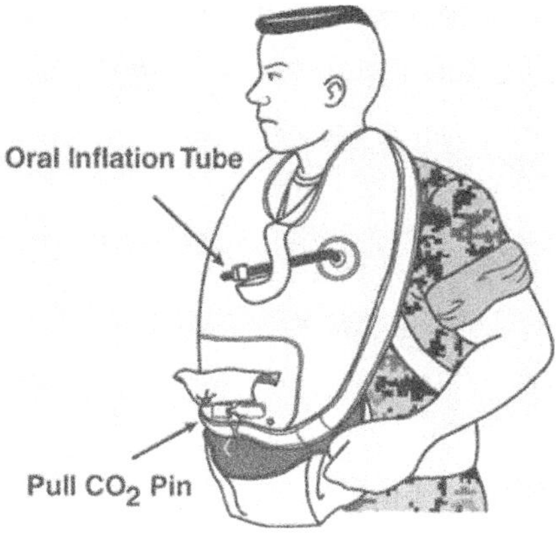

Staying Afloat Without a Life Preserver

You may be in open water without any floating objects or a life preserver to help you survive. If so, uniform blouses and trousers can be made into expedient flotation devices.

Floating With an Inflated Blouse

It is possible to float by a bubble of air trapped in the shoulders of your blouse. The air rises to the back and shoulders of the blouse and supports you at the water's surface. An inflated blouse is also a temporary flotation device used by weaker swimmers while trying to remove their trousers. There is a primary and an alternate way to create a bubble of trapped air in a blouse—

Primary Method
Turn the collar inside the blouse to help create a seal.

Unbutton top button and pull collar around mouth and nose.

Take a deep breath and bend forward slightly at the waist. Exhale one-half to three-quarters of a breath into the blouse.

Grasp and twist the collar with one hand to create a seal, this prevents air from escaping out from the collar.

Use your free hand and feet to stroke and kick to the surface.

Gather and hold the blouse tightly at the collar and stomach level to prevent the blouse from losing air if it floats up too high.

Splash water on the blouse periodically to prevent the material from drying, dry material allows air to escape.

Repeat inflation as required.

Alternate Method
Turn the collar inside the blouse to help create a seal.

Unbutton the second button from the top.

Take a deep breath and bend forward slightly at the waist.

Place your mouth and nose inside the hole created by the open button and exhale one-half to three-quarters of a breath into the blouse.

Grasp material at the unbutton portion and pull downward.

Use your free hand and feet to stroke and kick to the surface.

Splash water on the blouse periodically to prevent the material from drying, dry material allows air to escape.

Repeat inflation as required.

Floating With Inflated Trousers

In warm water, trousers can be used as a primary expedient flotation device. However, in cold water, submerging your head to remove and inflate your trousers results in heat and energy losses that negate the benefit of using the trousers as a flotation device. Once your trousers are inflated, you float motionlessly as if wearing a life preserver. If needed, assume the heat escape lessening posture (known as the HELP position, see page 1-37) to slow heat loss. As trousers dry, air leaks out of the legs. To slow this process, occasionally splash water on the fabric. Reinflate trousers as needed.

Sling Method

The sling method works if you are a strong swimmer or naturally very buoyant. Take the following steps to inflate trousers using the sling method:

Take a deep breath, bend over, and remove your boots.

NOTE: Retain your boots. Tie the boot laces together and suspend the boots from your blouse or hang them around your neck so that they rest on your chest.

Remove your trousers. Button or zip the trouser's fly closed. This allows you to control airflow.

Tie the bottoms of the trouser legs in a square knot. (The appendix illustrates various knots.)

Ensure that the front (fly) of the trousers faces you.

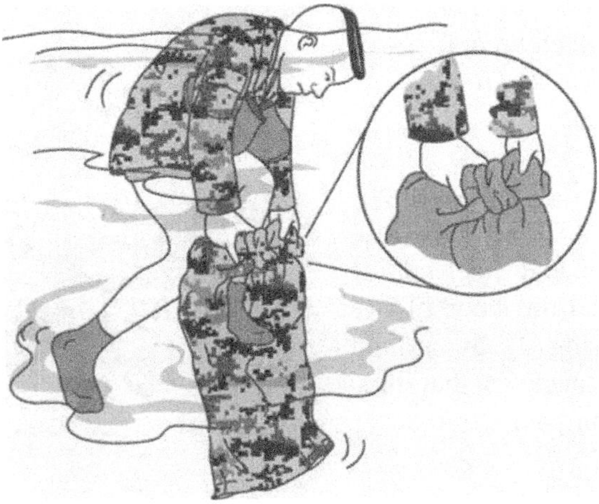

Hold the trousers above the water's surface and behind your head. Grasp both sides of the waistband and open with both hands.

Kick strongly to stay on top of the water while slinging the trousers overhead in order to trap air into them.

Once the waistband is submerged in the water, air is trapped in the legs.

Hold and seal the waistband underwater.

Slip the inflated legs over your head. Hold the waistband in toward your chest, the fly facing your body. To prevent air from escaping from the trousers, seal the waistband by either folding it or twisting it.

Lie back and relax, resting the back of your neck against the knot.

Splash water on the trousers periodically to prevent the material from drying. Dry material allows air to escape.

To replenish air in the trousers, you will use a technique known as the scooping method. With one hand on the open waistband, extend the trousers in front of you just below the surface of the water and scoop air bubbles with your free hand into the open waistband until the trousers have sufficient air. Repeat as necessary.

Splash Method

The splash method is an alternative to the sling method. As with the sling method, you must kick strongly to remain at the surface. To inflate trousers using the splash method, perform the following:

Take a deep breath, bend over, and remove your boots.

NOTE: Retain your boots. Tie the boot laces together and suspend the boots from your blouse or hang them around your neck so that they rest on your chest.

Remove your trousers. Button or zip the trouser fly closed. This allows you to control airflow.

Tie the bottoms of the trouser legs in a square knot. (The appendix illustrates various knots.)

Ensure that the front (fly) of the trousers faces you.

Hold the trousers at the water's surface out in front of you by the waistband with the fly up.

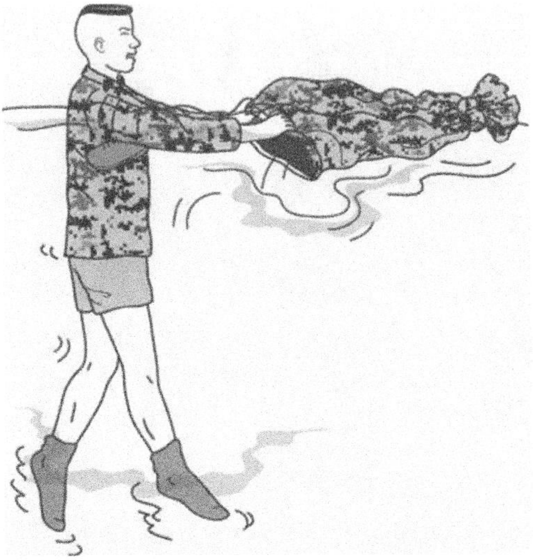

Grasp the waistband at the surface with one hand. Insert your free hand into the waistband, palm down.

Flutter your hand rapidly to create bubbles. This sends a mixture of water and air bubbles into the trousers. The water passes through the fabric. The air remains trapped in the legs.

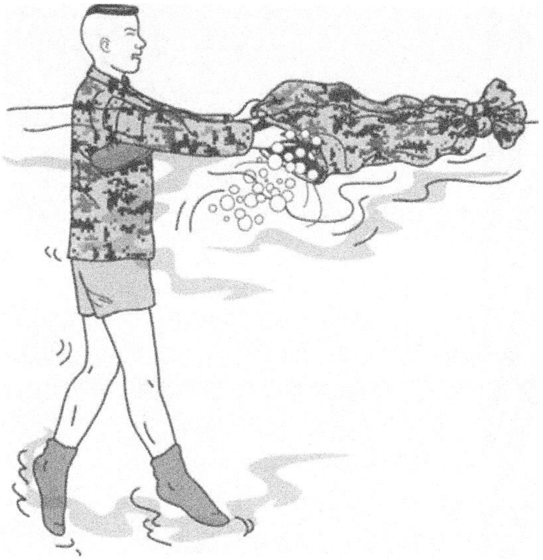

Hold and seal the waistband underwater.

Slip the inflated legs over your head. Hold the waistband in toward your chest, the fly facing your body. To prevent air from escaping from the trousers, seal the waistband by either folding it or twisting it.

Lie back and relax, resting the back of your neck against the knot.

Splash water on the trousers periodically to prevent the material from drying. Dry material allows air to escape.

To replenish air in the trousers, you will use a technique known as the scooping method. With one hand on the open waistband, extend the trousers in front of you just below the surface of the water and scoop air bubbles with your free hand into the open waistband until the trousers have sufficient air. Repeat as necessary.

Blow Method

The blow method is an alternative to the sling method. Use the blow method if you are a weak swimmer. Take the following steps to inflate trousers using the blow method:

Take a deep breath, bend over, and remove your boots.

NOTE: Retain your boots. Tie the boot laces together and suspend the boots from your blouse or hang them around your neck so that they rest on your chest.

Remove your trousers. Button or zip the trouser fly closed. This allows you to control airflow.

Tie the bottoms of the trouser legs in a square knot. (The appendix illustrates various knots.)

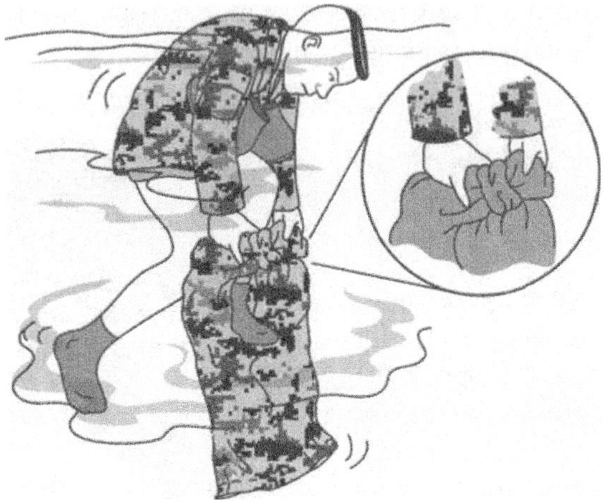

Ensure that the front (fly) of the trousers faces you.

Hold the trousers at the water's surface. Grasp both sides of the waistband and open with both hands.

Take a deep breath.

Drop 2 feet below the water's surface, pulling the waistband underwater.

Hold the waistband open with both hands and blow air into the trousers.

To fill the trousers with air, surface while keeping the waistband underwater, breathe in again, drop below the water's surface, and blow air into the trousers. Repeat these steps until the trousers are filled sufficiently. Once trousers are filled—

Hold the waistband underwater. Twist and pinch it off.

Slip the inflated legs over your head. Hold the waistband in toward your chest, the fly facing your body. To prevent air from escaping from the trousers, seal the waistband by either folding it or twisting it.

Lie back and relax, resting the back of your neck against the knot.

Splash water on the trousers periodically, to prevent the material from drying. Dry material allows air to escape.

To replenish air in the trousers, you will use a technique referred to as the scooping method. With one hand on the open waistband, extend the trousers in front of you just below the surface of the water and scoop air bubbles with your free hand into the open waist-band until the trousers have sufficient air. Repeat as necessary.

Avoiding Heat Loss in Cold Water

The rate of heat exchange in the water is about 25 times greater than it is in air of the same temperature. When you are immersed in cold water, hypothermia occurs rapidly due to the decreased insulating quality of wet clothing and as a result of water displacing the layer of still air that normally surrounds the body. You also lose about 50 percent of your body heat through your head; therefore, keep your head out of the water. Other areas of high heat loss are the neck, the armpits/sides, and the groin.

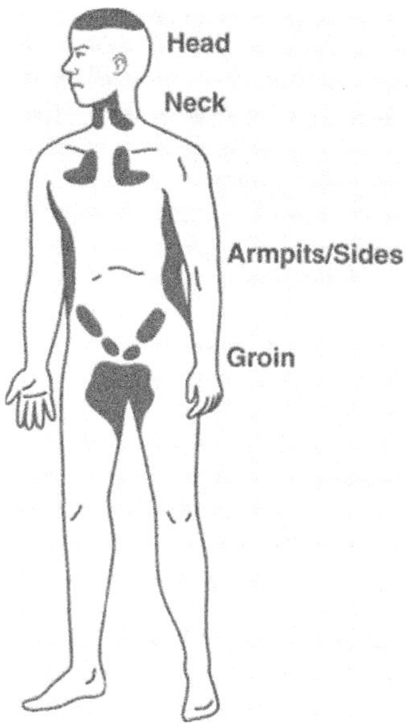

In cold water, DO NOT SWIM TO STAY WARM. Swimming, even with a slow and steady stroke, produces a lot of heat that is lost in the water. The heat loss can produce hypothermia that slows body functions and can result in serious injury or death. Remaining motionless conserves body heat three times longer than swimming. SWIM only if you have a flotation device and the shoreline is visible.

Individual Protection From the Cold

If you are equipped with a life preserver, assume the heat escape lessening posture (known as the HELP position) to slow heat loss and to protect major blood vessels near the body's surface. These areas lack insulating fat and are vulnerable to the chilling effects of cold water. To assume the HELP position—

Tuck your chin down tightly to cover your throat.

Draw your legs up in a fetal position to protect the groin.

Place your arms across your chest, tuck your hands into your armpits.

Wear some type of head covering (e.g., stowed cover, towel, handkerchief) to lessen heat loss through the scalp if head covering is available.

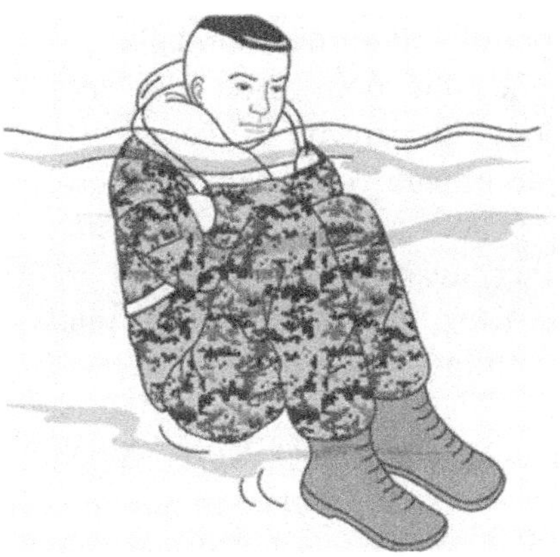

Group Protection From the Cold

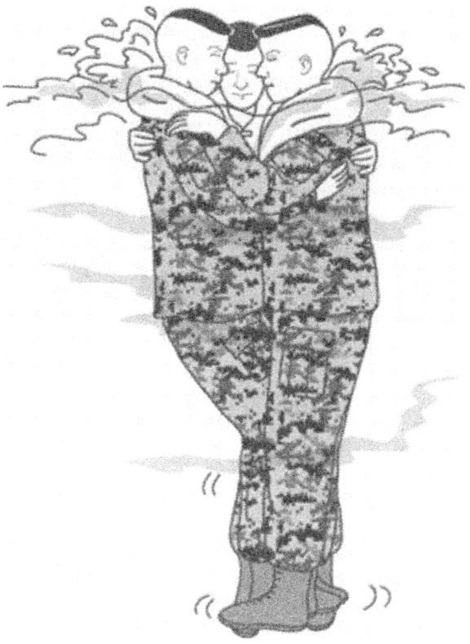

If three or more Marines are in the water and are equipped with life preservers, they should wedge tightly together and lock arms to form a circle known as a huddle position. This position protects vulnerable areas from heat loss.

A casualty who is suffering from the effects of the cold can be placed within the huddle to be surrounded by warmer water. If in the water for a prolonged period, it is recommended that everyone be rotated inside the huddle to maintain or re-warm each person's internal core temperature. If there are more than five Marines, they should make clusters of huddle positions.

Contact with other swimmers provides survival advantages—

Creates a larger target for search and rescue aircraft.

Provides additional warmth in cold water.

Improves morale.

Re-establishes the chain of command.

Reduces shock and panic.

Provides opportunities to administer first aid.

Supports exhausted Marines.

Drownproofing Methods

An object that floats on the surface has positive buoyancy. An object that floats a few feet beneath the surface has neutral buoyancy. An object that sinks has negative buoyancy. Most people have positive buoyancy and will float at the water's surface. To test your buoyancy—

Stand in water that is at your chest level or deeper.

Take a full breath.

Bend forward slowly.

Relax and wait.

If you have positive buoyancy, you will slowly rise to the surface. If you have neutral buoyancy, you will float a few feet beneath the surface. If you have negative buoyancy, you will sink. Regardless of whether you naturally float or sink, you can remain at the surface for extended periods without a life preserver if you exercise the appropriate drownproofing methods or survival strokes, which are based on your buoyancy.

Drownproofing methods consist of the T-method and the sweep. The breast stroke, side stroke, and elementary backstroke are the most common survival strokes. With any drownproofing method or survival stroke, remember the acronym **SAFE**:

SLOW EASY MOVEMENTS
Move slowly to conserve energy and minimize heat loss.

APPLY NATURAL BUOYANCY
Use natural buoyancy to support the body.

FULL LUNG INFLATION
Fill the lungs with each breath. Do not hold air in the cheeks.

EXTREME RELAXATION
Tight muscles are denser than relaxed ones and do not float as well.

Crawl Stroke

The crawl stroke, sometimes called the front crawl or free style, is the fastest stroke. To execute the crawl stroke—

Body position
Lie horizontal, on your stomach, in the water.

Look forward and downward at a 45 degree angle with the waterline between your eyebrows and hairline. Your head position is important as it assists in cutting a path through the

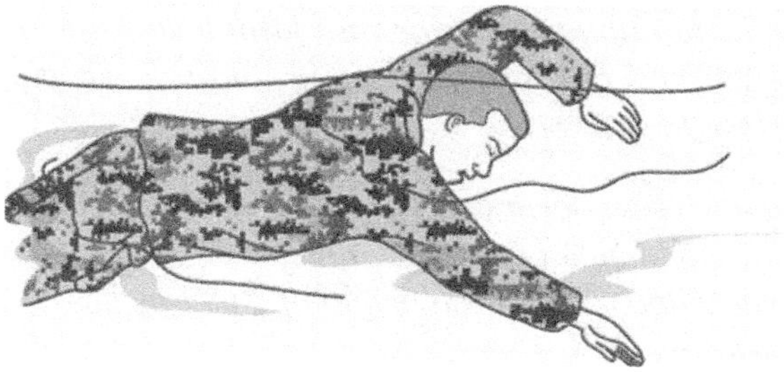

water. If your head is too high in the water, your lower body sinks significantly, making your stroke less efficient. If your head is too low in the water, water washes over your shoulders and neck, causing unnecessary drag.

Arm action
Arm action occurs in three phases: catch, propulsion, and recovery.

Fully extend one arm forward of your body, this positions your hand to catch the water in preparation for the propulsion phase.

To catch the water, bend your wrist (with your palm pointing outboard) and make an "S" shape (or inverted "S" shape) with your hand, ensuring that your hand does not cross the center of your body. Your left hand makes an "S" shape and your right hand makes an inverted "S" shape with your hand finishing at shoulder level.

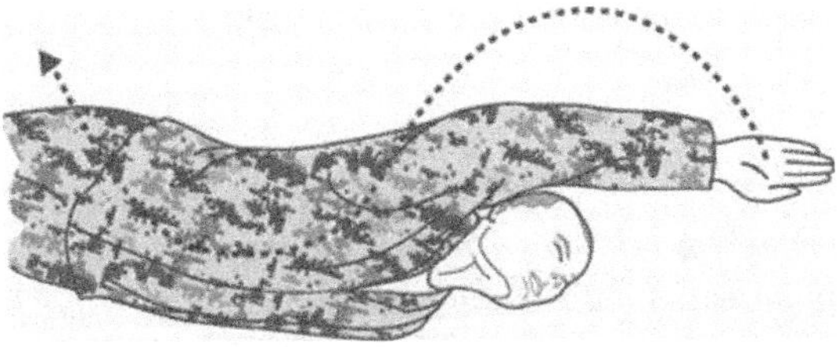

Push with your hands in a rearward fashion toward your feet until your arm is fully extended along your side, keeping your hands close to your body.

To begin the recovery phase, bend your arm at the elbow and raise your hand out of the water. Your hand breaks the surface of the water and maintains a height of 2 to 3 inches above the water's surface.

With your hand and arm moving in a forward manner, bring your hand past your head until your arm is about three-quarters of the way extended. At this point, turn your hand with your palm outboard so as to allow your thumb and fore-finger to enter the water first.

Once your hand enters the water, continue to push your arm forward until it is fully extended.

You are now prepared to catch the water again. These steps are performed in an alternating pattern: when one arm is catching and propelling, the other arm is recovering.

Leg action

Use a flutter kick to create the leg action for the crawl stroke. This kick is used for both propulsion and keeping the lower body horizontal with the water's surface. The flutter kick is an alternating leg action: one leg is kicking in a downward motion while the other leg is recovering to the surface to prepare for the next kick. The size of the kick ranges from 12 to 15 inches and depends on your height.

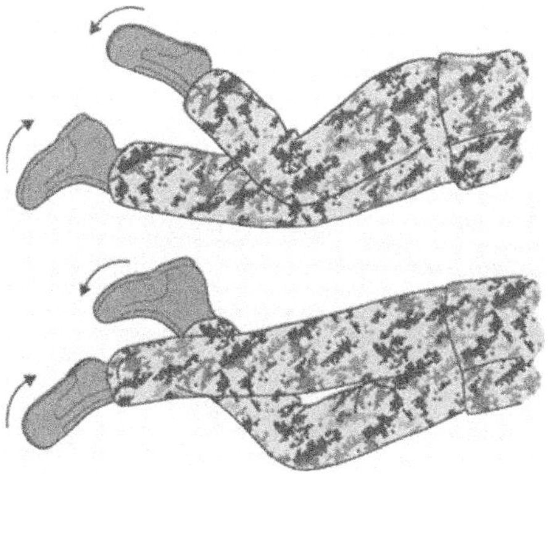

Maintain your legs in a semi-rigid manner.

Generate power for the kick from the hips.

Keep your feet loose at the ankles, they trail behind your legs and act as "flippers."

Execute the propulsion phase of this kick with a downward thrust of your leg.

Execute the recovery phase by pushing your leg back to the surface. This phase is complete when your foot reaches the surface.

Breathing
Breathe during the recovery phase of the arm action.

Roll your body and rotate your head to the side of your body where the arm recovery is occurring, this rolls the water away from your mouth. Keep your chin pushed back toward your shoulder.

Exhale while your face is still submerged and inhale when your face breaks the surface.

Breathe either bilaterally or rhythmically. To breathe rhythmically, breathe on the same side of your body every time your arm cycle occurs. Bilateral breathing requires you to breathe every one and a half arm cycles. To breathe bilaterally, breathe when your right arm is recovering, your face goes back in the water and the next time you breathe is when your left arm is starting its recovery three arm strokes later.

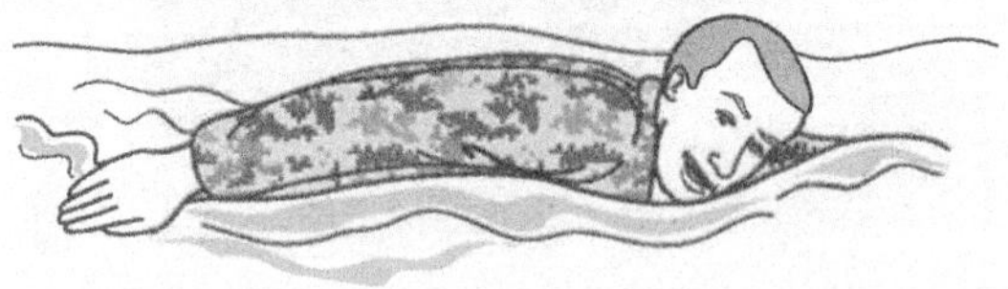

Bilateral breathing is the preferred method for breathing; it prevents the chance of hyperventilation and allows your body to maintain a lateral position in the direction you are swimming.

Coordination
This stroke uses constant arm and leg action.

T-Method

The T-method is a basic drownproofing method. This is the best survival technique if you have negative buoyancy. To execute the T-method—

With your face out of the water, take a deep breath and submerge your face in the water while holding your breath.

Float with your body in a horizontal position, arms extended from your side, and legs extended and joined.

Move your hands up to your armpits by tracing an imaginary line along your ribs.

Extend your arms outward (horizontal) to your sides, your body position resembles the letter "T."

Step out and forward with one leg and point your other leg to the rear, your knees should be slightly bent. Simultaneously, bring your arms down to your sides. Then bring your legs back together. You complete the steps by exhaling most of your air and preparing to surface your face to obtain another breath of air. Hold your head out of the water. Tilt your head back slightly. Breathe normally.

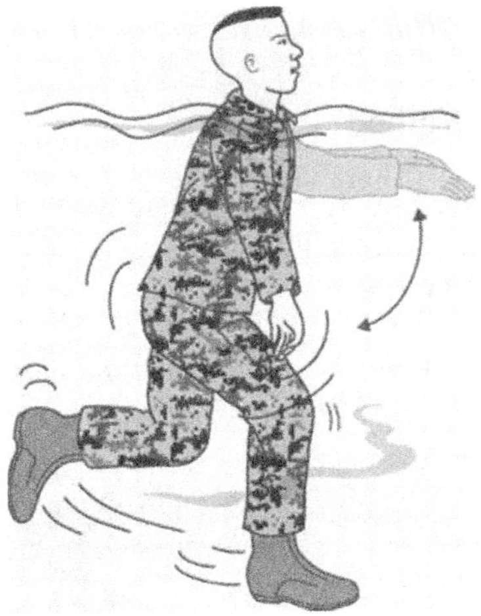

Once your breath is complete, move your hands up and down directly in front of your body. Do this two or three times to slow your descent into the water.

> NOTE: To avoid hyperventilating, hold your breath below the surface of the water for no more than 10 seconds.

The Sweep

The sweep works well if you have slight to excellent positive buoyancy. To execute the sweep—

Float face down in the water: bend 45 degrees at the waist, arms and legs dangling, head hanging down, relax all muscles.

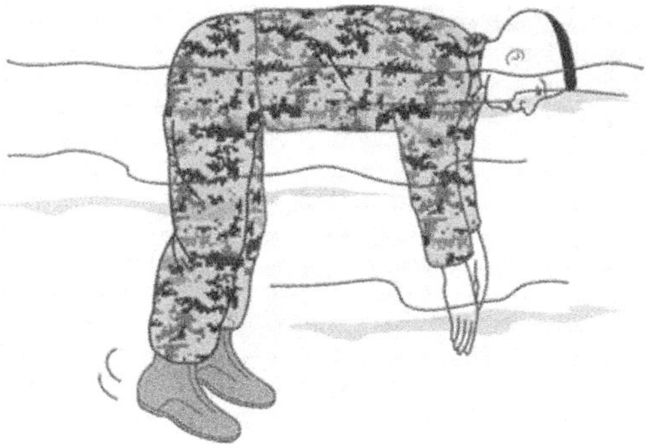

Spread your feet slowly to prepare for a single kick (one leg is forward and one leg is rearward).

Cross your arms in front of your chest, palms outboard with the back of each hand touching the opposite ear.

Exhale prior to raising your head for a breath.

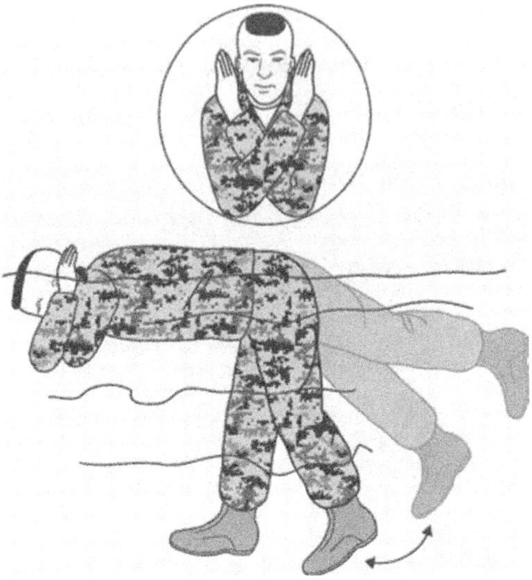

Bring your legs together and sweep your arms down and out until the arms are fully extended out to the sides. This raises your face above the water and allows you to catch a breath of air.

With your air supply replenished, return your face to the water and relax while sweeping your arms in a downward motion in front of your body to prevent/slow your descent. Do not hold your breath for more than 10 seconds. More than 10 seconds enhances your chances of shallow water black out and subsequent drowning.

Breast Stroke

Use this stroke to swim underwater, through oil or debris, and in rough seas. If you are a good swimmer and not wearing combat gear, the breast stroke is the best stroke for long-range swimming because it provides good visibility and allows you to conserve your energy and maintain a reasonable speed.

Body position
Lie prone in the water. Swim with your trunk and legs projecting back and down at an angle of 20 to 30 degrees.

Extend arms out in front (hands together [side by side]), and extend legs behind (toes pointed) to prevent drag.

Face downward, looking forward at a 45 degree angle to break the water and to prevent water from washing into the collar area causing drag.

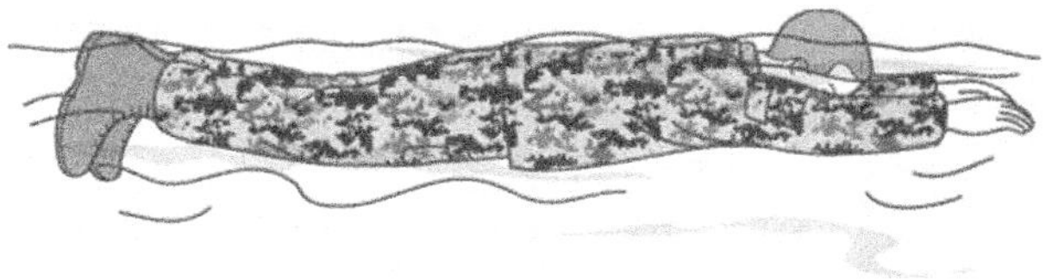

This is known as a glide.

Arm action

Turn your palms outward and bend your arms slightly.

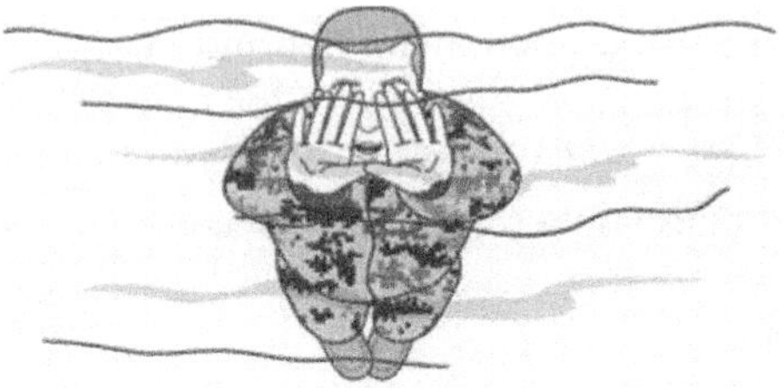

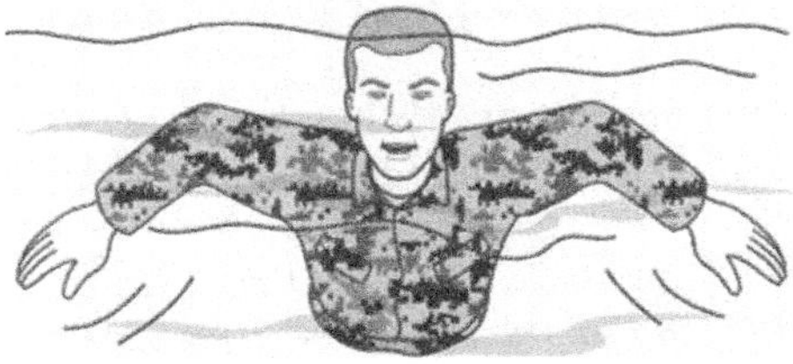

Sweep your arms sideward and slightly downward until your hands are opposite and slightly below your shoulders.

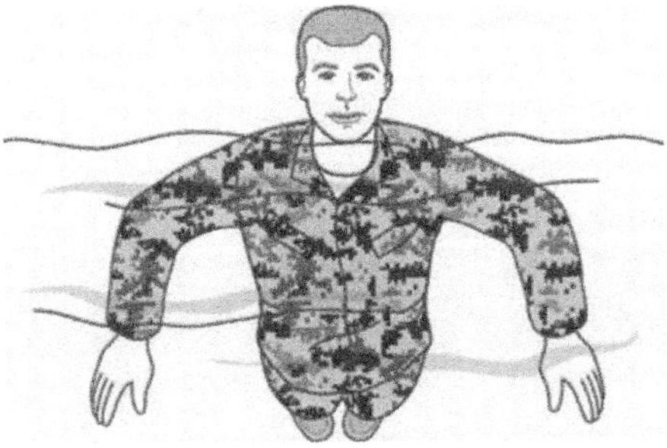

Rotate your head up, breathe once your mouth breaks the surface.

Bring your hands and arms up along your chest and thrust them forward until they are extended and ready to execute the next arm pull.

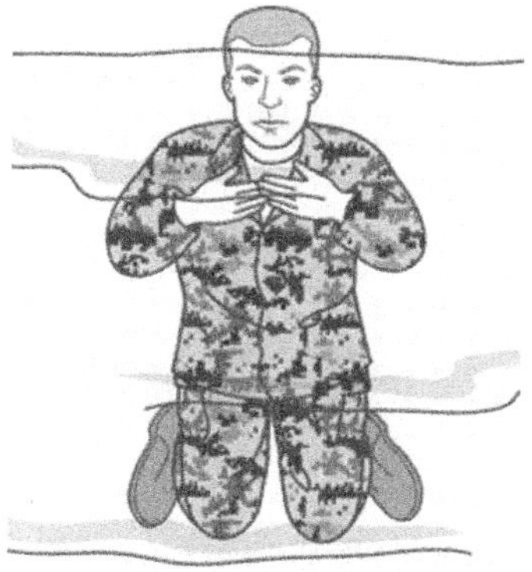

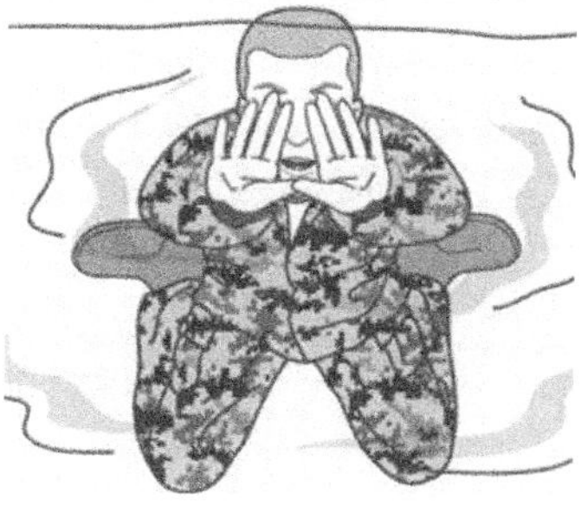

As the arms start their recovery into the glide, the head should rotate forward, resubmerging the face.

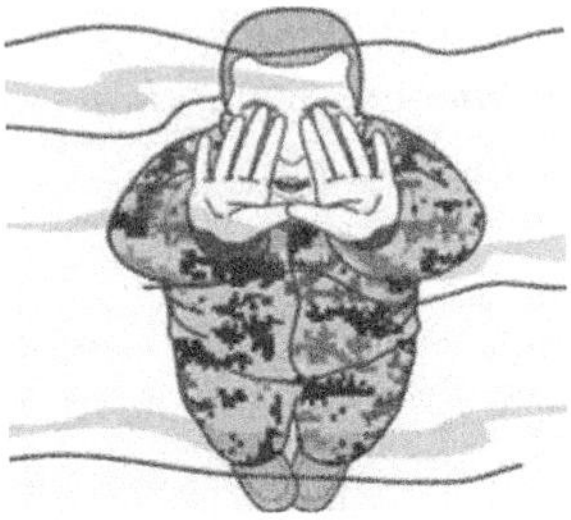

Leg action

Draw your heels toward your buttocks, establish a 45 degree bend in the knees.

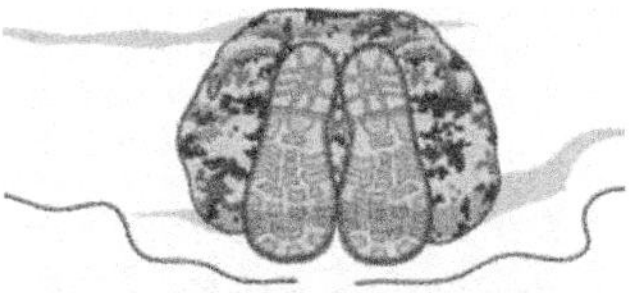

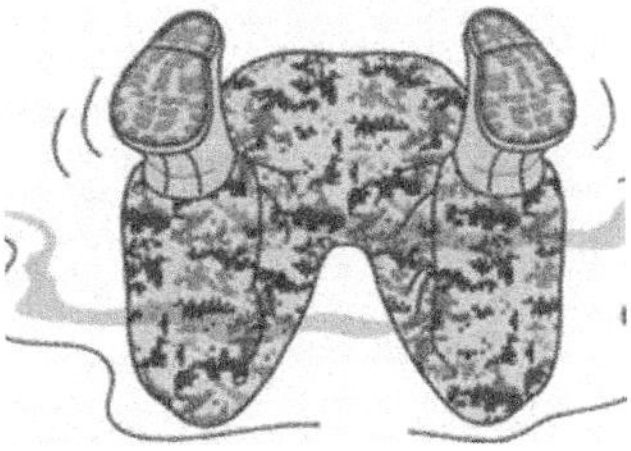

Thrust your legs outward and rearward, then squeeze them together. The whipping action of the feet aids forward propulsion.

This is known as the breast stroke kick.

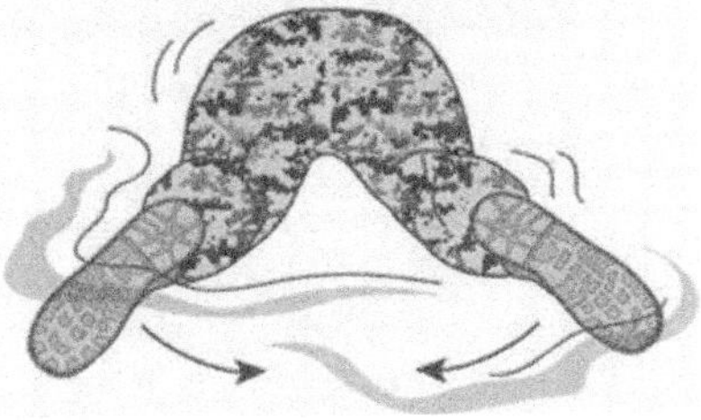

Breathing
Inhale during the arm pull and exhale through your mouth and nose during the finish of the breast stroke kick and glide.

Coordination
The stroke movement is in three counts:

Begin your arm pull. Near the finish of the pull, flex your knees and bring your heels toward your buttocks. The arm pull counteracts the resistance created by the knees.

As the arm pull is completed, thrust your hands forward, kick your legs outward and rearward, and squeeze them together.

Glide through the water for approximately 1 to 3 seconds or until your forward momentum decreases, then begin the next stroke.

Side Stroke

The side stroke is a survival stroke because you use both arms for buoyancy, with each arm creating a slight propulsion. The majority of your body's propulsion comes from your kick. To execute the side stroke—

Body position
Lie on your side with your lead (bottom) arm extended beyond (with a slight bend in your elbow) your head and in line with your body. Palm is down and your hand is submerged 6 to 8 inches.

Extend your trail (top) arm down the length of your body over your thigh.

Keep your legs straight and together, toes pointed rearward.

Keep your face out of the water, this allows for free breathing.

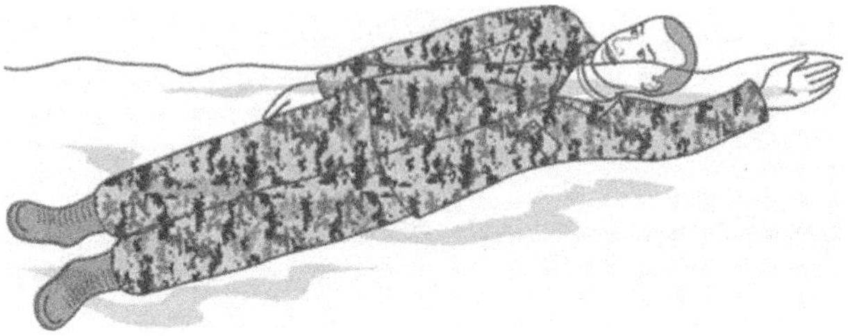

This is known as the glide.

Arm action
With your lead arm, pull your arm downward, while flexing at the elbow, until it is straight down from your shoulder.

Rotate your shoulder and pull your elbow into your side. This should put your lead hand at shoulder level. At the same time, turn your palm toward your face and thrust forward to your original, extended position.

Draw your right hand upward in front of your chest to shoulder level. Rotate your palm toward your feet, then push it downward in front of your body toward your feet to catch the water.

Push your trail hand backward to its original position on top of your thigh. (Your trail hand starts forward and meets your lead hand at your chest/shoulder.)

Leg action

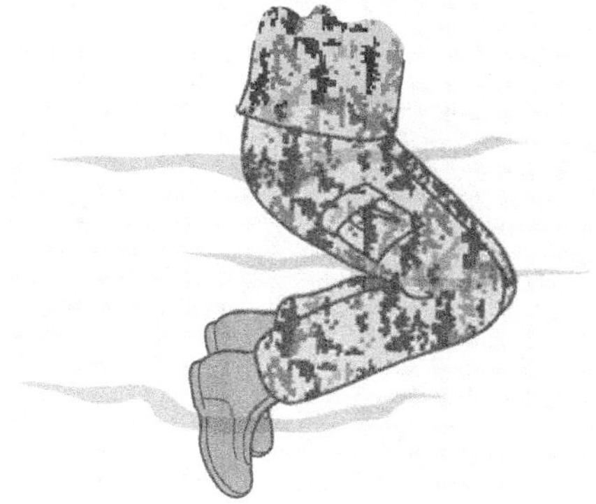

To perform the scissor kick, the top leg always goes forward and the bottom leg always goes rearward. From the extended position, draw—or recover—your feet toward your buttocks until your legs are bent at a 45 degree angle at the knees and the hips are flexed at a 45 degree angle with the thighs.

Once the legs have completed their recovery and while maintaining a 45 degree bend in the knees, extend the legs fully into a "V" shape in order to catch the water for the propulsion phase.

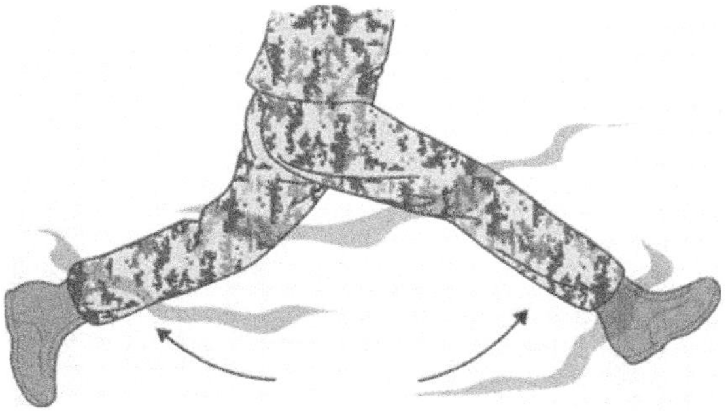

Once the legs are separated and extended forward and rear-ward to the "V" position, sweep the legs together until the feet are together.

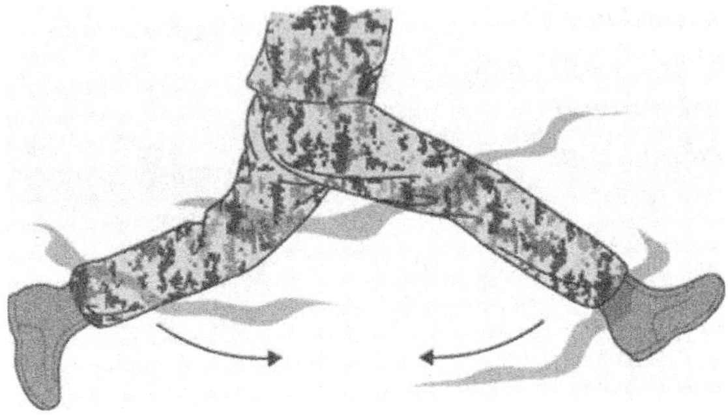

You are now in the glide position.

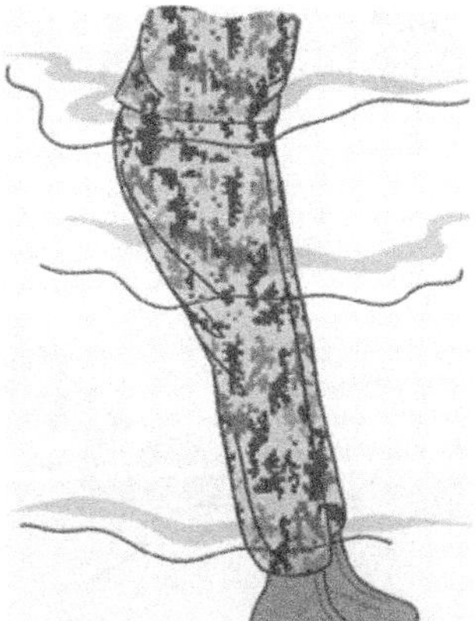

Breathing

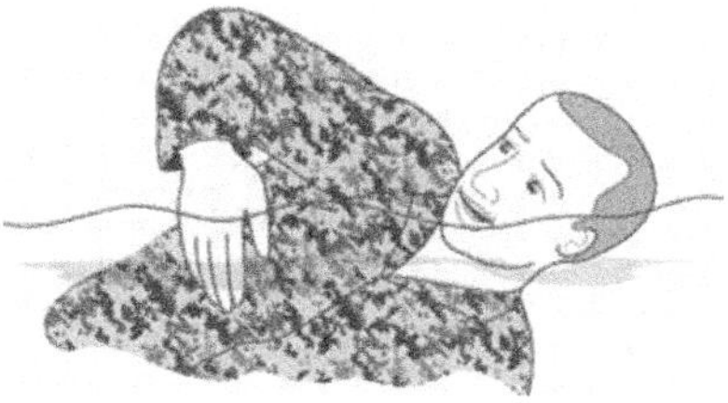

As long as your face remains clear of the water, it is a free breathing stroke. However, it is recommended that you exhale then quickly inhale when the legs are sweeping back together in the scissor kick. This is when the body reaches its highest point in the water, thus clearing the face completely from the surface of the water making it the optimum time to breathe.

Coordination

Begin the stroke with the downward pull of your lead arm. At the same time, bring your trail arm upward and draw your knees up to begin the kick. Let the thrust of the lead arm, push of your trail arm, and the kick of your legs coincide in order to finish the glide position. Glide through the water for approximately 1 to 3 seconds or until your forward momentum decreases, then begin the next stroke.

Elementary Backstroke

The elementary backstroke is also an excellent survival stroke. It relieves the muscles that you use for other strokes, and it is the recommended stroke for weak swimmers or nonswimmers. To execute the elementary backstroke—

Body position
Start on your back.

Face up, chest up, and hips up, keeping an arch in your lower back with arms pressed to your sides and your legs extended and joined to prevent drag.

Arm action
Trace your hands up your sides to an area near your armpits then extend your arms out to the sides to form the letter "T" (palms facing feet), locking out the elbows.

> NOTE: Don't raise your arms above your head. This creates drag, changes your body position, and submerges the head.

Slap your palms to your thighs using a strong sweeping motion.

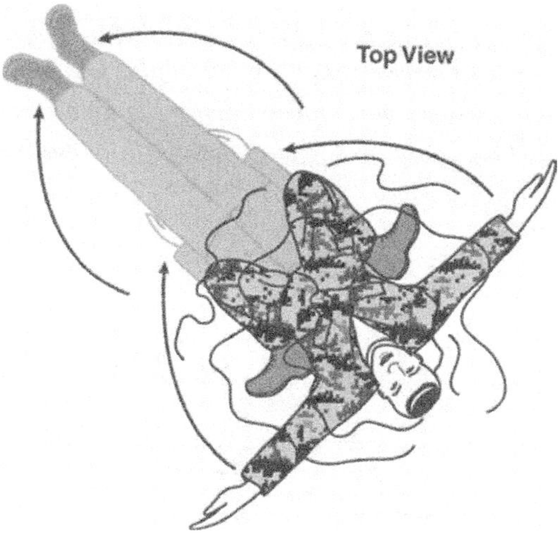

Leg action

Bend both legs at the knee (90 degree angle) slightly separating your knees and drawing your heels downward to a point under and outside your knees. The knees are spread as wide as the hips or slightly wider depending on the body type of the swimmer.

Circle around in a whipping action, ending with legs in a glide position.

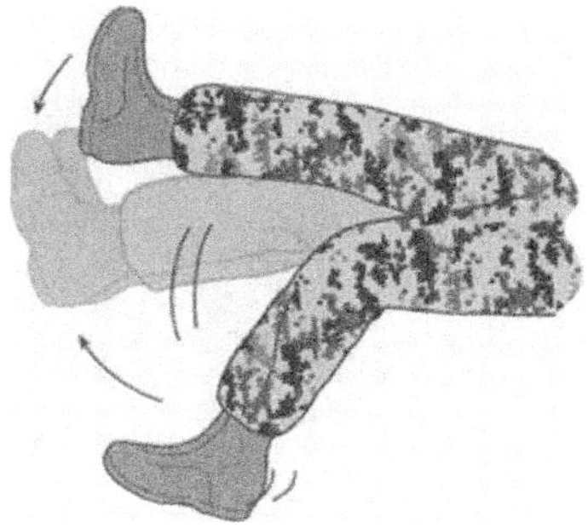

Breathing
Breathe anytime during this stroke. However, it is recommended that you exhale then quickly inhale when your arms are sweeping back toward your sides and while your legs are sweeping back together. This is when the body reaches its highest point in the water, thus clearing the face completely from the surface of the water and making it the optimum time to breathe.

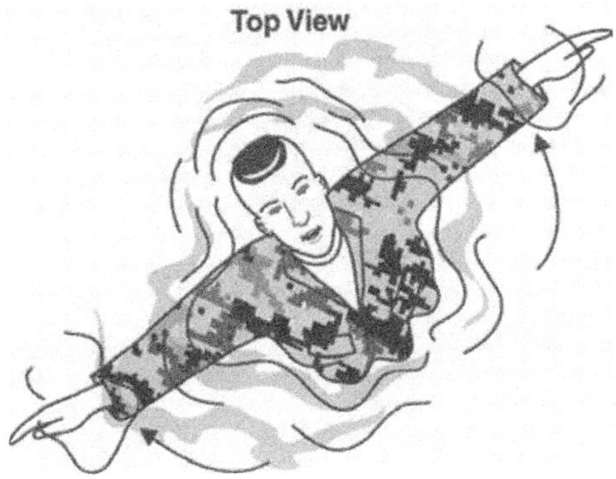

Coordination
The stroke movement occurs in three counts (recovery, catch, power).

Begin the arm pull (recovery).

Near the finish of the pull, flex your knees to a 90 degree angle. The arm pull counteracts the resistance created by the knees.

Kick out your legs, and squeeze them together as the arm pull is completed (catch, power).

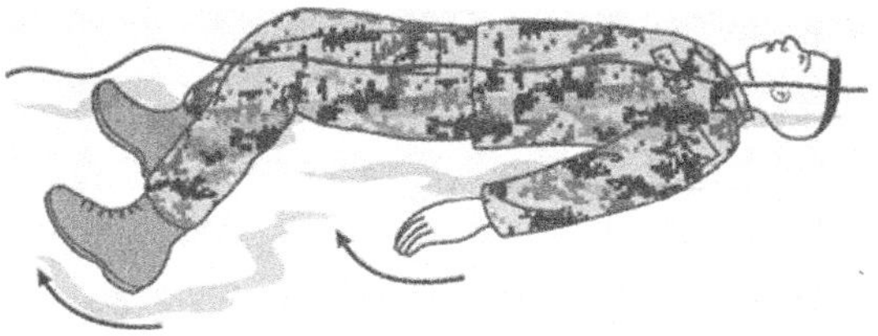

Glide through the water for 1 to 3 seconds or until your forward momentum decreases, then begin the next stroke as your momentum slows.

Chapter 2
Water Rescues

A drowning victim can panic and react with unexpected violence and can seize and inadvertently drown a rescuer. Therefore, if possible, a water rescue should be executed from a distance. Reaching, wading, or throwing methods are used in a distance rescue.

If a victim is too far away to use these methods, a swimming rescue may be necessary. Remove all combat gear, when possible, before entering the water. Swim within 2 to 6 yards of the victim to maintain a margin of safety, this allows you to reassess the situation and reassure the victim. If the victim is unconscious, use the wrist tow method or cross-chest carry method to pull the person to safety. If the victim is struggling, use a rear approach and then execute either a single armpit level off or a double armpit level off before towing the victim to safety.

If the victim does begin to overpower you, there are techniques that allow you to defend yourself without having to abandon the rescue. These techniques include the block, the wrist-grip escape, the front head-hold escape, and the rear head-hold escape. These techniques allow you to separate yourself from the victim, reassess the situation, and then attempt the rescue again.

> *Note:* In this chapter's illustrations, the rescuer is shown without a helmet.

Reaching Rescue Techniques

Reach

From a safe position at the water's edge, reach out to the victim. Talk constantly to calm the victim. Retain partial contact with land or some solid support structure (e.g., pier, bridge). If the victim is close but still beyond reach, extend an object (e.g., stick; pack; rifle with magazine removed, chamber empty, and muzzle pointing toward victim) that the victim can grasp. Pull the victim slowly to safety. Once the victim is close to shore, remove the victim from the water.

NOTE: You can also extend a foot to the victim if you can retain a secure grip on a solid support structure.

Reach From a Deck

Reach from a deck rescue can be executed by either a swimmer or a nonswimmer, and it can be used on an active or a passive victim. Be sure to reassure the victim during the rescue. To execute reach from a deck—

Lie prone on the deck with your body firmly anchored. To anchor your body, lie flat, spread legs apart, and extend one arm behind you with your palm down and on the deck.

Keep as much of your weight on the deck as possible and extend your free hand to the victim.

Grasp the victim's wrist from above, your thumb and index finger are facing you. The victim should never be allowed to grab you and put your life at risk. Therefore, never reach across the victim to grab his wrist, always grab the wrist that is the closest to you.

Keep your arm straight and locked out and pull the victim to the side of the deck. The victim should never be allowed to grab you and put your life at risk. Therefore, never pull the victim into you, always pull the victim into the side.

Arm Extension

An arm extension rescue can be used if you cannot reach the victim using the reach from a deck rescue technique and you must enter the water. An arm extension rescue technique is used for a victim who is either active or passive. Once you determine that a reach from a deck rescue technique is not viable—

Reassure the victim and quickly ease into the water while holding onto the deck with one hand.

Grasp the victim's wrist from above, your thumb and index finger are facing you. The victim should never be allowed to grab you and put your life at risk. Therefore, never reach across the victim to grab his wrist, always grab the wrist that is the closest to you.

Keep your arm straight and locked out and pull the victim to your side. The victim should never be allowed to grab you and put your life at risk. Therefore, never pull the victim into you, always pull the victim into the side.

Leg Extension

If the victim is beyond the reach of your arm, ensure that you have a firm grip on the deck, extend a leg to the victim and allow him to grab it, slowly bring the victim in closer until you can grab the victim's wrist that is holding onto your leg. Once you have a firm grasp on the victim's wrist, use the steps in the arm extension to pull the victim to safety. The leg extension rescue can only be used to rescue an active victim because the victim must be able to grab the rescuer's leg.

Wading Assist

Do not wade into water that is deeper than your chest. Talk constantly to calm the victim. If possible, do not touch the victim directly. Extend an object (e.g., stick; pack; rifle with magazine removed, chamber empty, and muzzle pointing toward victim) that the victim can grasp. Once the victim grasps the object, pull the victim slowly to safety.

Throw

If the victim is not within reach, use an expedient line to throw a lifesaving device to the victim. A lifesaving device can be any weighted item that floats (e.g., a canteen that is one fourth full of water). The lifesaving device must be secured to the end of the rope so that the rope will feed out from its coil when tossed. Talk constantly to calm the victim. Once the victim grasps the line or the lifesaving device, pull the victim toward you at a steady pace that keeps the victim's head above the water's surface. DO NOT pull so strongly that you break the victim's grip on the line. To prepare and use an expedient line and lifesaving device—

Tie a bowline at one end of the rope. (The appendix illustrates various types of knots.)

Unfasten the lid of a canteen.

Place the bowline around the neck of the canteen.

Refasten the lid so the canteen hangs from the bowline loop.

Place one end of the rope under the ball of your forward foot to secure it, either tie a knot at the end of the rope or tie an object at the end of the rope to create a block. Stand with your weight on the end of the rope. DO NOT tie the rope around your ankle.

Coil 20 to 30 yards of the rope, and hold it in a nonthrowing hand.

Place the canteen in your throwing hand.

Use an underhand throw to pitch the canteen and rope a short distance over the victim's head. Keep your nonthrowing hand open so the coil can unfold freely. The rope should trail across the victim's outstretched arms.

Retrieve the rope if the throw is inaccurate or the victim fails to grasp it. Recoil the rope as it is retrieved.

Divide the coil and throw again.

Lifesaving Approaches

Properly approaching the victim is as important as any other aspect of a rescue. Knowing the appropriate and safest method needed to approach a victim is critical in ensuring your safety and the victim's survival. Determining the victim's physical state (distressed swimmer, victim that is active and drowning, or victim that is passive and drowning) is crucial and will determine which type of approach you will execute.

Front Surface Approach

The front surface approach is typically performed when the victim is passive. You must remember that approaching the victim from the front is dangerous because a distressed swimmer or victim that is active can lunge toward you.

Rear Approach

If the victim is active, you must remember that his response can change rapidly. Therefore, approaching him from the front could be extremely dangerous. The victim could easily grab you and take you under the water during a state of panic. Therefore, use a rear approach when possible. When approaching a victim from the rear, reassure the victim until you get to within 2 to 6 yards of the victim; at that point, stop talking to the victim in order to maintain an element of surprise.

Approach Strokes

Before approaching a victim, you must evaluate the victim:

- Is the victim a distressed swimmer?
- Is the victim active and drowning?
- Is the victim passive and drowning?

Select your approach stroke based on your evaluation. Remember, the safety of the rescuer is paramount. Do not endanger yourself in an attempt to reach a drowning victim.

Crawl Stroke Approach Stroke

The crawl stroke (free style) is the fastest approach stroke. This stroke is used when the victim is passive and/or unconscious in the water. That is, the victim is face down, submerged or near the surface of the water, is not breathing, or is not moving. If any of these conditions exist, there is an obvious need to reach the victim as quickly as possible.

Execution of the crawl stroke approach stroke is the same as the crawl stroke (see pages 1-40 through 1-44), except that your head remains above the water's surface to allow for free breathing and maintaining eye contact with the victim.

Maintaining eye contact is critical. If the victim becomes submerged, you will have a better chance of locating him if you know his last location. Your distance from the victim will determine if your head remains out of the water the entire time you are executing the stroke. If you are 54 yards or less from the victim, you should maintain eye contact by keeping your head raised until you reach the victim. If you must swim more than 54 yards to reach the victim, make eye contact with the victim, place your face back in the water for four or five strokes, then raise your head again to regain eye contact with the victim (repeat these steps until you reach the victim).

Body position
The position of your body is horizontal to diagonal and influenced by the position of the head. This allows you to maintain eye contact with the victim.

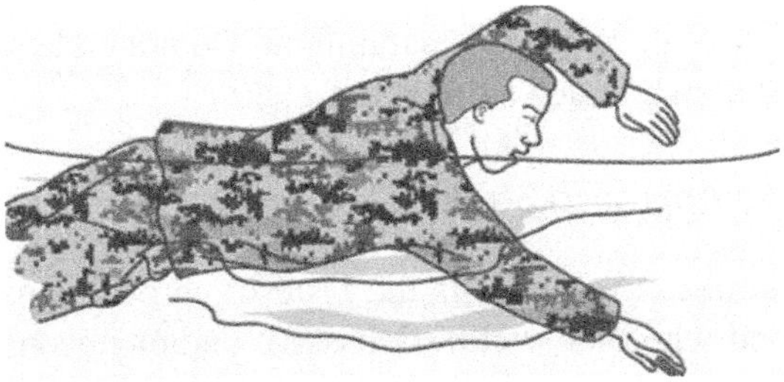

Arm action
While one arm is forward of your head to catch the water for the propulsion phase, extend your other arm alongside your body in what is known as the recovery phase. The recovery phase ends when your arm is extended forward of your head to begin the propulsion phase. In the propulsion phase, your arms are approximately shoulder-width apart and the pulling action is slightly wider and deeper to compensate for your raised head.

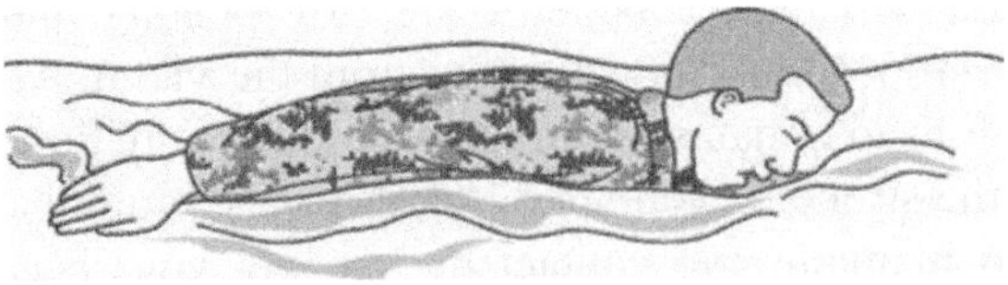

Leg action

Kick one leg in a downward motion while your other leg is recovering to the surface to prepare for the next kick (this is known as a flutter kick). This kick is used for propulsion and maintaining the lower body as horizontal to the water's surface as possible. When your head is raised, your knees are slightly bent in order to keep your feet near the surface.

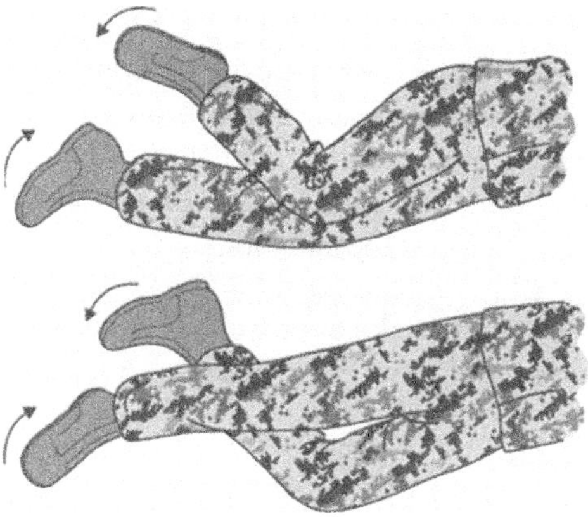

Breathing

If your face is above the water, breathe as needed (free breathing).

If traveling in excess of 54 yards to reach the victim, rotate your head to the right or left during the recovery phase and inhale when your face is out of the water or wait until your head is out of the water and facing forward while regaining sight of the victim.

Coordination

This stroke uses constant arm and leg action.

Breast Stroke Approach Stroke

The breast stroke approach stroke is used when a swimmer is distressed or a victim is active and drowning. This stroke is also used when the victim has a suspected spinal injury because it minimizes the wake created around the victim, minimizes the movement of the victim's head and neck, and helps prevent further injury.

If a victim is at a distance greater than 54 yards and if the swimmer is in distress or the victim is active and drowning in open water (ocean, lake), you should use the crawl stroke approach stroke to rapidly approach the victim, stopping approximately 11 yards from the victim in order to assess the situation.

Execution of the breast stroke approach stroke is the same as the breast stroke with a few minor deviations.

Body position
The position of your body is horizontal to diagonal and influenced by the position of the head. This allows you to maintain eye contact with the victim. It also allows you to communicate with the victim in order reassure him as you approach.

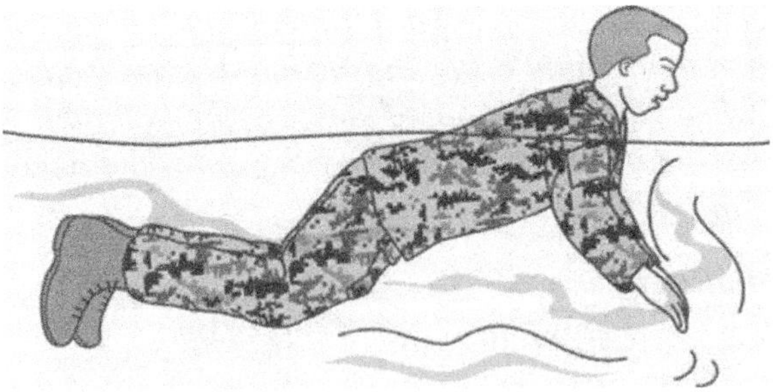

Arm action

Execute the arm action for the breast stroke, except that your arm pull is wider and deeper, which allows your head to remain above the water's surface.

Leg action
Execute the leg action for the breast stroke, except that your legs open wider during the propulsion phase of the stroke.

Breathing
Keep your face above the water and breathe as needed (free breathing).

Coordination
Coordination is the same as the breast stroke, except that there is no glide period. You constantly stroke until you reach the victim.

Level Offs

A properly performed level off positions the victim's face up and horizontal to the water's surface. There are three types of level offs: front surface approach, single armpit level off, and the double armpit level off. The single armpit level off and the double armpit level off are done by performing a rear approach, which is the preferred approach when a victim is active.

Front Surface Approach

The front surface approach is performed when the victim is passive. Approaching a victim from the front may place you in danger because a distressed swimmer or a victim who is active may lunge toward you. To execute a front surface approach—

Stop 2 to 4 yards from the victim to reassess the situation and reassure the victim.

Determine which arm to use in order to rotate the victim into a face-up position (your right hand on the victim's right wrist or your left hand on the victim's left wrist).

Move into position and prepare to grasp the victim.

Turn sideways and move toward the front of victim.

Once in position, reach forward and grab the victim's wrist. Your thumb is on the inside of the victim's wrist, as if you were checking his pulse with your thumb. Your remaining fingers wrap around the victim's wrist.

Lean back immediately and execute a powerful scissor kick or inverted scissor kick and perform short, vigorous pulls with your free arm.

As the victim begins to move forward in the water, kick, pull, and twist outboard on the victim's wrist. The momentum created from the kick and the pulling and twisting action of the victim's wrist will rotate the victim into a face-up and horizontal position in the water.

Extend and lock out your towing arm down the length of your body and execute a wrist tow to move the victim to safety.

Single Armpit Level Off

Approach the victim.

Stop 2 to 6 yards from the victim to reassess the situation, reassure the victim, and maintain your margin of safety.

Approach the victim slowly and grab his armpit with your hand (your right hand to victim's right armpit or your left hand to victim's left armpit). Position yourself sideways to the victim and place your elbow in the center of his back.

Pull with your hand that is in the victim's armpit while pushing with your elbow that is in the victim's back to place the victim horizontal on the water's surface (face up). To assist in placing the victim horizontally, use your free arm to execute short, vigorous pulls and use your legs to execute a scissor kick or inverted scissor kick.

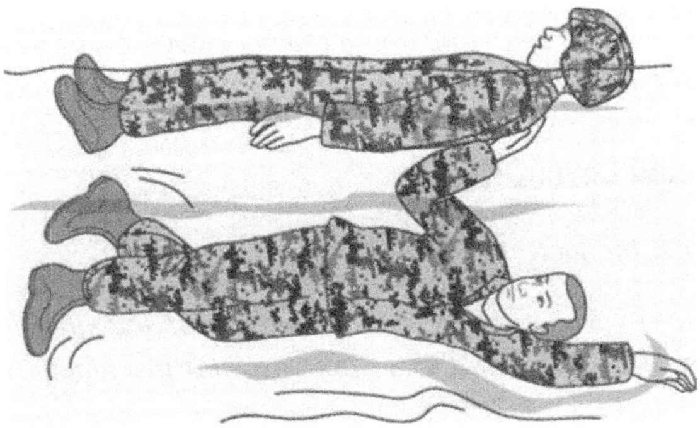

Once the victim is horizontal and on the water's surface, begin forward momentum by extending your towing arm to a fully locked-out position and executing a single armpit tow.

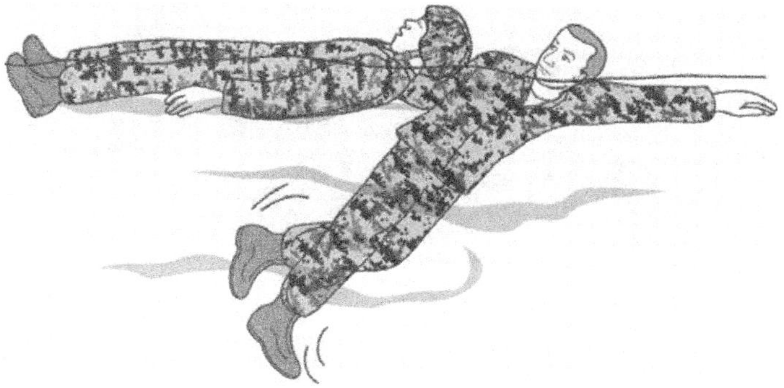

Double Armpit Level Off

Approach the victim.

Stop 2 to 6 yards from the victim to reassess the situation, reassure the victim, and maintain your margin of safety.

Approach the victim slowly and grab his armpits with both your hands. Place your elbows on his back.

Pull with your hands that are in the victim's armpits while pushing with your

elbows that are in the victim's back and use an inverted breast stroke kick to place the victim horizontal on the water's surface (face up).

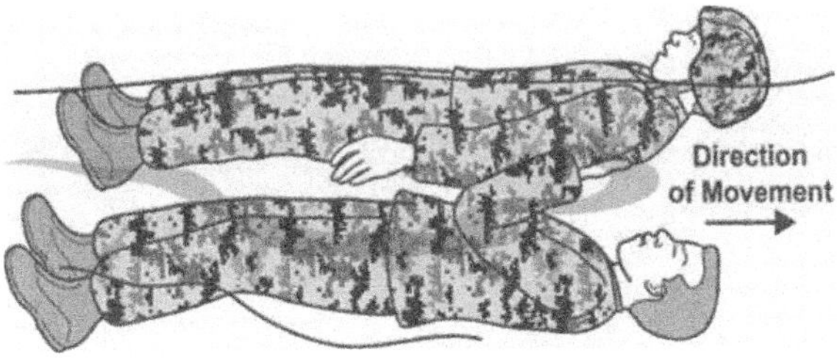

Once the victim is horizontal and on the water's surface, begin rearward momentum by slowly extending your arms into a fully locked-out position and executing a double armpit tow and inverted breast stroke kick.

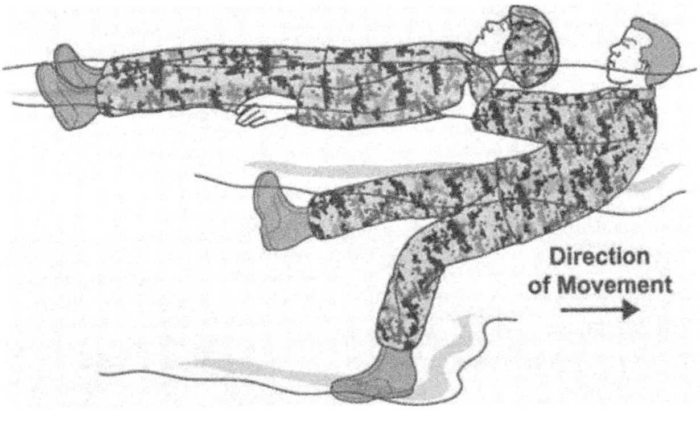

Rescue Techniques

Wrist Tow

Use the wrist tow method to rescue a victim who is floating face down. DO NOT use the wrist tow on a struggling victim. If time allows, remove your helmet and gear before attempting the rescue. Swim toward the victim using a modified breast stroke. Swim within 2 to 6 yards of the victim to maintain a margin of safety, this allows you to reassess the situation and reassure the victim. The following steps show proper front surface approach, wrist tow procedures:

Approach the victim from the front and grasp the underside of the victim's left wrist with your left hand or the right wrist with your right hand. Ensure that your thumb is on the underside of victim's wrist.

Lean back, pulling and kicking strongly to move the victim into a horizontal position.

Twist the victim's wrist to rotate the victim into a face-up position.

Swim toward safety using the lifesaving stroke.

> NOTE: The lifesaving stroke is a modified side stroke wherein the top arm is used to tow or carry a victim to safety and use a scissor kick or an inverted scissor kick is used for propulsion.

Keep a firm grip on the victim's wrist.

Keep your towing arm fully extended and along your side. This keeps the victim in the water column and prevents drag.

Ensure the victim's head does not go under water during the recovery.

Single Armpit Tow

You perform the single armpit tow after a victim has been properly leveled off. The single armpit tow uses the lifesaving stroke. To execute the single armpit tow—

Place one hand (top arm) in the victim's armpit. Your towing arm is straight and along your side.

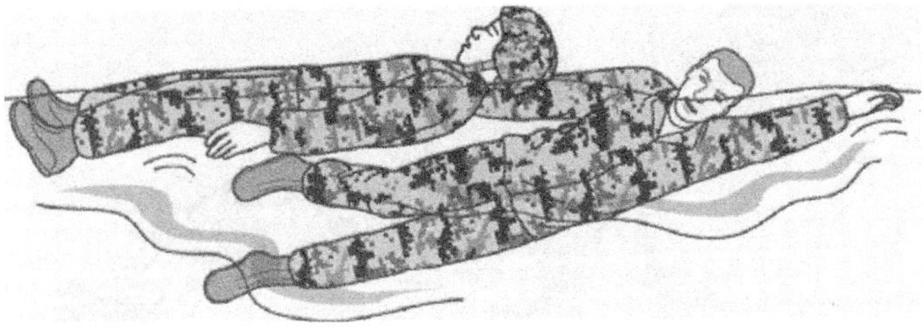

Use your lead arm (bottom arm) to execute short, vigorous pulls.

Execute a scissor or inverted scissor kick in a continuous and vigorous manner.

Use either free breathing or explosive breathing during the tow. During free breathing, keep your head above the water's surface and continuously reassure the victim. During

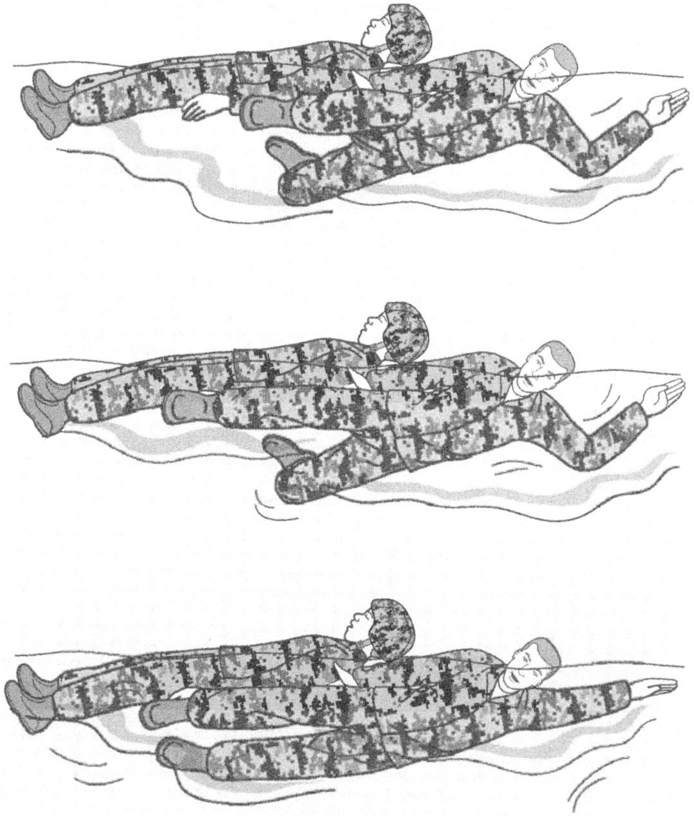

explosive breathing, put your head under the water to plane off your body's angle and reduce drag. Execute approximately two strokes, then lift your head up to breath and reassure the victim. Return your head back underwater for the next few strokes.

Double Armpit Tow

You perform the double armpit tow after a victim has been properly leveled off. To execute the double armpit tow—

Place both hands in the victim's armpits.

Extend both arms fully along your body. You are on your back, and your face is clear of the water.

Use free breathing.

Reassure the victim at all times.

Use the inverted breast stroke kick (it is the only kick that can be used due to your body position) to tow the victim to safety. (The inverted breast stroke kick is the same kick used in the elementary backstroke.) Your kick must be continuous and vigorous in order to keep the victim's face above the water.

Collar Tow

You perform the collar tow after a victim has been properly leveled off. To perform a collar tow using the victim's blouse—

Maintain control of the victim by grasping his armpit with one hand and then with your free hand grasp either his combat gear or his blouse between his shoulder blades. If grasping the blouse, grasp the material with your palm up, then turn your hand over to tighten the material.

Release the victim's armpit once control is established and execute the lifesaving stroke to tow the him to safety.

Cross-Chest Carry

Use the cross-chest carry to carry a victim to safety if the victim is struggling or when moving through heavy surf. Remove your helmet and gear before attempting a rescue. Talk constantly to calm the victim. Swim toward the victim using a breast stroke approach stroke. Swim within 2 to 6 yards of the victim to maintain a margin of safety, this allows you to reassess the situation and reassure the victim. The following steps illustrate proper cross-chest carry procedures:

> **CAUTION:** The cross-chest carry causes fatigue even if you are in excellent physical condition.

Use a level-off technique to place the victim in a horizontal, face-up position.

Retain a grip on the victim with one hand. Reach over the victim with your free hand to encircle the victim's chest. Place your free hand on the victim's opposite rib cage, just below his armpit.

Release your grip once you have a secure hold on the victim's chest.

Swim toward safety using the lifesaving stroke while keeping a firm grip on the victim's chest and your hip on his back.

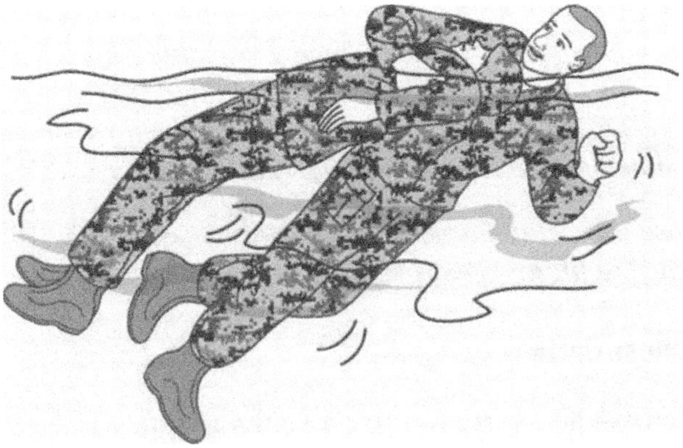

This procedure brings the victim's face and shoulders clear of the water, and typically the victim stops struggling. Sometimes, however, the victim will struggle during the swim to safety. If this happens, either tighten your grip on the victim or defend yourself with one of the techniques discussed on pages 2-25 through 2-30.

Tired Swimmer's Assist

You must maintain a 2 to 6 yards margin of safety from the victim at all times while you are getting into position to perform the tired swimmer's assist. Once behind the victim, you extend one arm, hand straight. Place your straight hand underneath the victim's opposite armpit. Maintain a 45 degree angle away from the victim, your arm is locked out. Assist the victim in propulsion until both of you reach safety.

Defense Against a Drowning Victim

DO NOT sacrifice your life in an attempt to save the victim. A struggling drowning victim poses great danger to anyone nearby. Driven by panic, the victim can grab you with great strength in an effort to climb out of the water. This can result in death for both you and the victim. The following techniques can be used to defend against a drowning victim, but the best defense against attack by the victim is to stay out of reach.

Block

The block prevents the victim from grabbing you if you have approached from the front. If the victim lunges toward you, react as follows:

Place one or both open hands against the victim's upper chest; being careful to avoid the victim's face, neck, and abdomen.

Lean backwards and submerge rapidly. Keep your blocking arm(s) extended.

Swim underwater and away from the victim, and return quickly to the surface.

Stop 2 to 6 yards from the victim to reassess the situation.

Determine an appropriate course of action.

Wrist-Grip Escape/Wrist-Grip Escape Alternative

The wrist-grip escape is used when a victim grabs your arm or wrist. Submerge the victim quickly by reaching across with your free hand, pushing down on the victim's shoulder to submerge him, and kicking to propel yourself upward. While keeping the victim submerged with your hand on his shoulder, give three hard jerking pulls with your trapped hand in an attempt to break free from his grasp. Once free, swim clear of the victim and reassess the victim's condition.

If the wrist-grip escape is unsuccessful in freeing your hand, use the wrist-grip escape alternative: use your free hand to grab your trapped fist, rotate thumbs up, apply bone-on-bone contact with the victim's arm, pry your hand out of his grasp, and quickly swim away from the victim.

Front Head-Hold Escape

The front head-hold escape allows you to escape when you are facing a victim who is gripping you around your head and neck. To execute the front-head hold escape—

Take a quick breath and tuck your chin into your shoulder to protect your throat.

Clap your hands above your head (three times) to submerge instantly. This drags the victim below the water, lifts his arms from around your neck, and, typically, he releases his grasp in order to get back to the surface. If he doesn't release his grasp, apply pressure to the victim's brachial pressure points (which are located inside of the upper arm, above the elbow).

Thrust the victim's arms up and away.

Keep your chin tucked to protect your throat, and swim underwater away from the victim and return quickly to the surface at the ready position.

Stop 2 to 6 yards from the victim to reassess the situation.

Determine an appropriate course of action.

Rear Head-Hold Escape

The rear head-hold escape allows you to escape when a victim is gripping your head and neck from the rear. To execute the rear head-hold escape—

Take a quick breath. and tuck your chin down, turn your head to either side, and raise your shoulders to protect your throat.

Take a strong stroke, clap your hands above your head (three times), and submerge instantly. This drags the victim below the water and, typically, he releases his grasp in order to get back to the surface. If he doesn't release his grasp, apply pressure to the victim's brachial pressure points (which are located inside of the upper arm, above the elbow).

Thrust the victim's arms up and away.

Twist your head and shoulders until free.

Swim underwater away from the victim and return quickly to the surface.

Stop 2 to 6 yards from the victim to assess the situation.

Determine an appropriate course of action.

Administering First Aid/Rescue Breathing

To administer first aid/rescue breathing, remove the unconscious victim from the water when possible and stop any bleeding.

First, open the airway and check for breathing. If the victim is not breathing, give two breaths of air. Check for a pulse. If there is a pulse but the victim is not breathing, continue rescue breathing. If there is no pulse, start cardiopulmonary resuscitation (CPR). For detailed information on administering CPR and rescue breathing, refer to American Red Cross CPR skill cards.

WARNING

If the victim has no pulse and is not breathing, administer CPR immediately. If the victim does have a pulse but is not breathing, give rescue breathing only. If the victim has a pulse and is breathing, DO NOT give CPR—CPR could prove fatal.

Next, protect any wounds from exposure, and treat for shock.

Chapter 3
Treatment of Casualties and Avoidance of Dangerous Marine Life

To survive in the water, you will face many challenges other than just being able to swim. For example, injuries and/or fatigue can lead to drowning and exposure to the elements can lead to hypothermia or heat injuries and may require medical attention. Finally, your presence in the water may be seen as a threat to or food source for a variety of marine life and you may be attacked. This chapter advises you on how to minimize these risks in order to stay alive.

Drowning

Drowning is suffocation by a liquid. A drowning victim inhales water into the lungs or the throat closes by reflex so that little or no water can enter the windpipe. In either case, a victim can no longer breathe.

Symptoms

One symptom of drowning is that the victim may call for help and has an expression of dread or panic. But typically a victim that is active and drowning may not call for help because he is trying to conserve his air and will not speak.

Another symptom of drowning is when the victim thrashes at the water's surface. If the victim stops or grows calmer, he has likely been overcome by fatigue, hypothermia, or a lack of air. At this stage, the victim usually has 20 to 60 seconds before going under the water's surface.

Treatment

If the victim is not breathing, begin rescue breathing. Place the victim on his back, tilt head back to open airway, pinch the nose, and give two full breaths. If the victim does not inhale during the first two breaths, reposition his head and attempt two more breaths. Check for a pulse. If a pulse is present, but the victim is still not breathing, continue rescue breathing. If a pulse is not present, begin CPR. See MCRP 3-02G, *First Aid*, for rescue breathing and CPR details.

WARNING

If the victim has no pulse and is not breathing, administer CPR immediately. If the victim does have a pulse but is not breathing, give rescue breathing only. If the victim has a pulse and is breathing, DO NOT give CPR—CPR could prove fatal. Continue first aid until medical help arrives.

A victim who is not breathing and has no pulse may appear dead. DO NOT decide that death has occurred. Continue with the prescribed treatment. A corpsman or medical officer should decide whether the victim can or cannot be revived.

Hypothermia

Hypothermia is the abnormal lowering of the body's internal (or core) temperature to 95 degrees or below. It occurs when the body loses heat faster than it produces heat. A hypothermia victim loses the ability to move quickly, becomes mentally sluggish, slips into semiconsciousness, lapses into a coma, and dies when internal body temperatures drop too low.

The chilling effects of cold air, wind, or water can produce hypothermia. Water poses the greatest threat because it transfers heat 25 times faster than air. Depending on the water's temperature, a victim can succumb to hypothermia within a few minutes. The body's sudden contact with cold water can also set off a body reaction known as the mammalian diving reflex. This reflex can greatly increase survival time (especially for women and children) in or under cold water. The mammalian diving reflex shuts off blood circulation, except for the flow between the heart, lungs, and brain. The small amount of oxygen left in the blood and lungs is saved for the body's vital organs. This reflex has allowed people to survive being under cold water for an extended period of time. Therefore, a cold water drowning victim should be treated as if still alive even though the victim is not breathing, has no pulse, and may appear dead. DO NOT decide that death has occurred. Continue with the prescribed treatment. Victims of hypothermia can appear to be dead when they are not. A corpsman or medical officer should decide whether the victim can or cannot be revived.

WARNING

If the victim has no pulse and is not breathing, administer CPR immediately. If the victim does have a pulse but is not breathing, give rescue breathing only. If the victim has a pulse and is breathing, DO NOT give CPR—CPR could prove fatal. Continue first aid until medical help arrives. Check for a pulse for at least 45 seconds.

Symptoms

Once the body's core temperature drops, the victim will show one or more of the following symptoms:

- Violent and uncontrollable shivering as the body tries to warm itself.
- Slow or slurred speech.
- Disorientation or poor coordination.
- Loss of skin color.
- Blue and pinched lips.
- A slowing or stopping of shivering that progresses into a rigid torso and limbs.

Survival Time

A hypothermia victim's survival depends on the water's temperature and the time spent in the water. A small body build cools faster than a large build. Children cool faster than adults.

To increase your chance of survival in the water, utilize the HELP position described on page 1-37. Extra clothing and inactivity (remaining motionless in the water) can also increase your survival time.

Treatment

A hypothermia victim must be warmed to prevent further heat loss; therefore, treatment should begin as soon as possible. Consciousness of the victim determines the treatment that should be pursued. See MCRP 3-02G for specific treatment information. The following treatment procedures are recommended:

- If the victim is conscious, give the victim warm fluids. Give candy or sweetened foods to a victim who is able to eat.
- If the victim is unconscious, place him on his back with his head tilted back to ensure an open airway.
- DO NOT massage the victim. Massage can break blood vessels and create swelling, internal pressure, and blocked blood circulation.
- DO NOT give alcohol to the victim. Alcohol lowers the victim's body temperature.
- Shock is a possibility, treat accordingly.
- Seek medical help immediately.

If you are able to remove the victim from the water, apply the following steps when possible:

- Get the victim into shelter.
- Remove the victim's wet clothing.
- Put the victim in dry clothing.
- Place the victim in a sleeping bag if one is available. It may be necessary to place another Marine in the sleeping bag with the victim.
- Place as much insulation as possible between the victim and the ground.
- Use hot water bottles, electric blankets, or blankets heated in an oven or by a campfire to warm the victim's neck, groin, and the sides of the chest.

CAUTION: DO NOT apply heat to extremities.

Heat-Related Injuries

Heat-related injuries include heat cramps, heat exhaustion, and/or heat stroke. They can occur when a Marine—

- Is exposed to extreme heat, such as from the sun or a combination of high air temperatures and water temperatures.
- Does not wear proper clothing or gets overexposed to the sun.
- Becomes dehydrated.

With heat cramps, muscles may cramp in the arms, legs, and/or stomach. The victim may sweat excessively. To treat, create improvised shade for the victim and have him drink water.

With heat exhaustion, the victim sweats heavily; presents pale, moist, cool skin; and complains of a headache, weakness, dizziness, and/or loss of appetite. A victim may also experience heat cramps, nausea, vomiting, an urge to defecate, chills, rapid breathing, confusion, and a tingling sensation in the hands or feet. To treat, pour water over the victim and fan him to speed up the coolant effect of evaporation, have the victim drink freshwater, and attempt to provide shade.

With heat stroke, the victim stops sweating, which results in red, flushed, hot, dry skin. The victim may first experience headaches, dizziness, nausea, fast pulse and respiration, and/or seizures and mental confusion. The victim can become unconscious and die if not treated quickly. To treat, create improvised shade for the victim. Have him drink water and elevate both legs if possible.

Burns

Apply a field dressing (or the cleanest material available) to the burn. Give sips of water to a casualty who is conscious and not nauseated. When treating burn casualties—

- DO NOT remove clothing stuck to the burns.
- DO NOT break any blisters.
- DO NOT apply grease or ointments to the burns.

For electrical burns, check for both an entry and exit burn from the passage of electricity through the body. An exit burn may appear on any area of the body, not necessarily opposite the entry burn.

For burns caused by wet or dry chemicals, flush the burns with large amounts of water and cover with a dry dressing.

For burns caused by white phosphorus, flush the area with water, then cover with a wet material, dressing, or mud to exclude air and keep the white phosphorus particles from burning.

For laser burns, apply a field dressing.

Common Medical Problems Associated with Sea Survival

Seasickness

Seasickness is the nausea and vomiting caused by the bobbing motion created by the wave action of a flotation device. Seasickness can result in—

- Dehydration and exhaustion.
- A loss of the will to survive.
- Others becoming seasick.
- Unclean conditions.

To treat seasickness—

- Wash both the Marine and the flotation device to remove the sight and odor of vomit.
- Keep the Marine from eating food until the nausea is gone.
- Have the Marine lie down and rest.
- Give the Marine seasickness pills if available. If the Marine is unable to take them orally, the pills should be inserted rectally for absorption by the body.

Saltwater Sores

Saltwater sores occur when skin that has abrasions or is cut is exposed to saltwater. The sores may form scabs and pus. Do not open or drain. Flush the sores with freshwater, if available, and allow to dry. Apply antiseptic, if available.

Blindness/Headache

Irritants or the effects of the sun's rays reflecting off the water can cause temporary blindness or headaches. If flames, smoke, or other irritants get in your eyes, flush the eyes immediately with saltwater, then with freshwater, if available. Apply an ointment, if available. Bandage both eyes for 18 to 24 hours or longer if the damage is severe. If glare from the sky and water causes your eyes to become bloodshot and inflamed, bandage the eyes lightly. Try to prevent this problem by wearing sunglasses or goggles with a sunglass insert.

Constipation

This condition is a common problem associated with dehydration. For constipation, do not take a laxative if it is available; this causes further dehydration. Drink freshwater, if available.

Sunburn and Dehydration

The sun's rays reflect at all angles off the waves of the water; therefore, sunburn and dehydration are serious problems in sea survival. Try to prevent sunburn by—

- Erecting an improvised canopy, with available floating materials, to provide shade.
- Wearing your soft cover or using a cloth, such as a handkerchief, to cover your head.
- Covering your skin with sunscreen or lip balm from your first aid kit. Your lips, nostrils, eyelids, the backs of your ears, and the skin under your chin sunburns easily. If enough sunscreen cream is available, all exposed skin should be covered.

Dehydration is caused by the loss of the body's vital fluids. Dehydration in saltwater may result from a combination of factors such as a lack of water, the effects of saltwater on skin tissue, sunburn or vomiting from seasickness, and other causes. Sleep and rest and reduced water and food intake are the best ways of enduring periods of exposure. The following measures will delay the effects of dehydration:

- DO NOT drink saltwater.
- DO NOT drink urine.
- DO NOT drink alcohol.
- DO NOT smoke.
- DO NOT EAT unless water is available.

Dangerous Marine Life

You may see many types of marine life around you, and some are more dangerous than others. Generally, sharks are your greatest danger, followed by barracudas. However, most marine life will not deliberately attack a human. The most common injuries from marine life are wounds from bites, stings, or punctures. With the exception of sharks and barracudas, most injuries are a result of either trying to catch game or from contact abrasion with marine life. To treat an injury that results from a dangerous marine animal bite—

- Pack the wound with gauze and then apply a pressure bandage, if available.
- Treat for shock.
- Obtain medical attention as soon as possible.

For a more detailed information on the treatment of casualties and dangerous marine life, see MCRP 3-02F and MCRP 3-02G.

Sharks and Barracudas

Only about 20 percent of all shark species are known to attack people. Sharks have an acute sense of smell, and the smell of blood in the water will draw them to their prey. They are also very sensitive to any abnormal vibrations in the water; therefore, the sound caused by a struggling swimmer or underwater explosions will attract them.

Barracudas are bold and inquisitive fish. They have been known to attack men who are wearing shiny objects. Barracudas may charge at lights or shiny objects at night.

A group of swimmers can maintain a 360 degree watch while in the water. Therefore, to protect yourself from sharks and barracudas, stay with other swimmers. The group members can either frighten away or fight off sharks or barracudas better than one person can. Keep all your clothing on, including footwear. Historically, sharks attack unclothed individuals in groups first, mainly in the feet. Clothing also protects you against abrasions from a shark's tough skin should the shark brush up against you. Avoid urinating heavily, let urine dissipate between discharges. If you must defecate, do so in small amounts and toss it as far away from the group as possible. Do the same if you must vomit.

If attacked, the use of firearms by swim sentries should be used with extreme caution because of the risk of injury to other swimmers. If unarmed or unable to make an improvised weapon, kick and strike the shark. Avoid using your bare hands to strike the shark, injury can result to your hands due to a shark's tough skin. Target areas on a shark are the gills, eyes, and underbelly. Blows to the snout are also not recommended because a shark will tilt its head up and thrust its jaws forward when biting.

Sea Snakes

Sea snakes are venomous and sometimes found in mid-ocean. They are unlikely to bite unless provoked. Your best protective measure is to avoid them.

Poisonous Fish

Many reef fish are toxic, and can kill you if eaten. Generally, poisons are present in all parts of the fish, but especially concentrated in the liver, intestines, and eggs.

Turtles and Moray Eels

Turtles and moray eels normally inflict minor bite wounds. Treat this type of injury by cleaning the wound.

Corals

Coral, dead or alive, can inflict painful cuts. Clean all coral cuts thoroughly. DO NOT use iodine to disinfect any coral cuts. Some coral polyps feed on iodine and may grow inside the flesh.

Jellyfish, Portuguese Man-of-War, Anemones, and Others

This group of marine animals inflicts injury by stinging their victims with their tentacles. Contact with their tentacles produces burning pain, a rash, and small hemorrhages on the skin. Shock, muscular cramping, nausea, vomiting, and respiratory distress may also occur. Gently remove the clinging tentacles with a towel and wash or treat the area. Use diluted ammonia or alcohol and talcum powder to treat the injury if available.

Spiny Fish, Urchins, Stingrays, and Cone Shells

These animals inject their venom by puncturing the skin with their spines. General signs and symptoms include swelling, nausea, vomiting, generalized cramps, diarrhea, muscular paralysis, and shock. Deaths are rare. Treatment consists of soaking wounds in hot water, if available, to deactivate heat-sensitive toxins.

Chapter 4
Negotiating Water Obstacles

Marines face water obstacles in saltwater, freshwater, and brackish water (where freshwater and saltwater meet). These water environments differ considerably and pose distinct problems for Marine tactical units and swimmers. Saltwater obstacles include tides, surfs, and currents. Brackish water obstacles include back bays. Freshwater obstacles include rivers and canals.

Tides

Tides are periodic changes in the surface levels of oceans, bays, gulfs, inlets, and rivers. The Moon's and Sun's gravitational pulls cause tides. Depending on the situation, tides can either help or hinder Marines in their endeavors to conduct amphibious or riverine operations. Tides can provide sufficient depths to allow for the passage of landing craft or boats over reefs, trees, and other underwater obstructions. On the other hand, tides can render a river fording site unusable to tactical vehicles. Direction, level of change, and amount of change determine tidal nomenclature.

Tides that show change in direction are flood tides and ebb tides. Rising tides are known as flood tides. Falling tides are known as ebb tides.

Tides that show extreme levels of change are high tides and low tides. High tide is the period when water is at its greatest depth. Low tide is the period when water is at its most shallow depth.

Tides that show the least and most amount of change are neap tides and spring tides, respectively. Neap tides have the least amount of change in water levels between high and low tide. Neap tides occur at the half moon, when the Sun and the Moon are aligned at a 90-degree angle with the Earth. In this position, the Sun's and Moon's gravitational pulls offset each other. Spring tides have the highest floods and lowest ebbs. Spring tides occur at or shortly after the new moon or full moon when the Sun, Moon, and Earth are approximately in line. In this position, the Sun's and Moon's gravitational pulls are combined.

Surf

Waves break upon entering shallow water and create surf. The offshore area where waves break is the surf zone, which can present many hazards. Breaking waves often trap air bubbles and create a foamy appearance. Bubbles lower the water's density and decrease buoyancy. Move through foamy surf as quickly as possible. The type of wave determines your survival technique.

Wave action moves you toward shore. Lie on your back or side with your head pointing in the direction of the beach and your feet pointing into the waves.

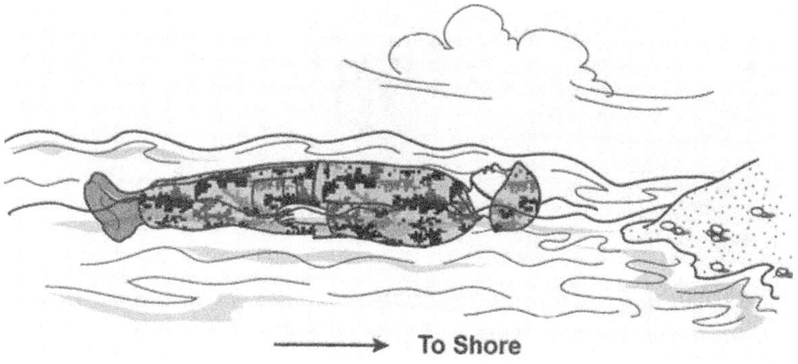

As one wave approaches the beach, another drains away; relax and do not swim against the draining water. When a new wave is within about 3 yards, start swimming toward shore. Continue to swim until the wave lifts you and moves you toward the beach. Once the wave loses forward momentum, relax and repeat the cycle. If nearing rocks, turn your body and approach feet first to reduce the chance of striking your head and arms.

Plunging Waves

A plunging wave is a breaker whose crest curls forward and falls ahead of its base. Because of its power and underwater turbulence, a plunging wave poses the greatest surf threat. If caught in a plunging wave, you can be pulled underwater and pitched about violently. This can cause you to panic, which can increase your chance of drowning. Perform the following steps to escape a plunging wave:

Tuck into a ball with your head against your knees and your forearms locked around your legs, just below the knees.

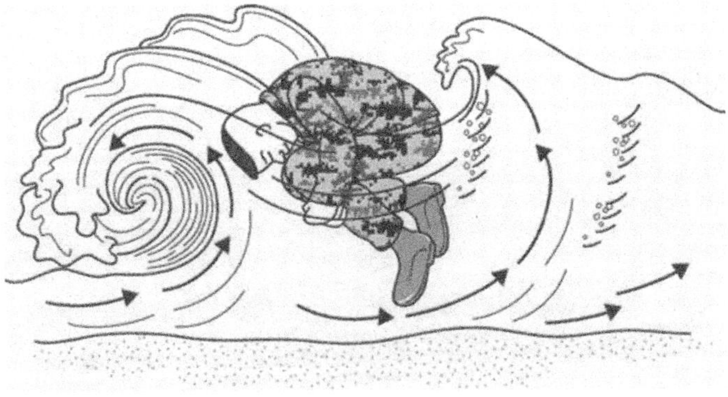

Relax in this position until the turbulence subsides and you float to the surface. This can take 30 seconds or more.

Swim toward shore.

NOTE: If threatened by another plunging wave, dive underwater into the wave.

Spilling Waves

A spilling wave does not break. Instead, its crest slides forward without curling. A spilling wave creates less turbulence and poses less of a threat than a plunging wave. If caught in a spilling wave, relax and let the wave carry you to shore.

Surging Waves

A surging wave occurs on a beach with a steep underwater gradient. It never really breaks, but the crest rises while the base slides up the beach with great force and speed. Once the wave reaches its highest point on the beach, it rushes back as quickly as it surged forward. If you are standing on the bottom when a surging wave advances or retreats, the wave can knock you off your feet and pull you into the surf zone. Do not try to stand or walk on the bottom. Swim toward the beach as soon as possible.

Currents

Offshore Currents

An offshore current occurs outside the surf zone. Typically, it occurs at bay entrances, in island channels, and between islands and the mainland. An offshore current flows parallel to or away from shore. If the offshore current is created by tides, its current strength and direction vary at different times of the day.

If caught in an offshore current, you may be carried in a direction you do not want to go. DO NOT try to swim directly to safety. If the current is moving directly away from the shore, relax and wait until the current dies out or turns toward land. Once the current subsides, use any survival stroke to swim toward shore. If the current is moving parallel to shore, use any stroke to move at an angle across the current and toward shore.

Rip Currents

A rip current occurs when waves pile water against the shore faster than the water can drain. The water flows rapidly along the beach until it is deflected seaward by a bottom obstruction. Then the rip current flows through the surf zone and into open water at a speed of up to 2 knots. This action can cut deep trenches in the sand. A rip current dies out once it reaches open water (usually within a few hundred yards of the shore).

A rip current can pull you out to the open sea. If caught in a rip current, DO NOT try to swim against the current. A rip current moves faster than most people swim, and it is impossible to swim to shore once caught in it. Relax and stay afloat until the current runs out. Once the current subsides, use a survival stroke to move parallel to the shore until you are out of the current, then begin swimming toward shore.

Littoral Currents

A littoral current occurs when a wave breaks against a beach at an angle. This current flows parallel to the shoreline and does not pose a great threat. If caught in a littoral current, use the combat travel stroke to swim across it at an oblique angle.

Back Bays

Once on the beach, you often face one or more rows of low hills called dunes. Behind the dunes, you may encounter a low-lying stretch of ground thickly covered with scrub trees and bushes. This area gives way to wetlands known as back bays. Back bays consist of muddy islands that are almost submerged during flood tides and separated by channels of brackish water of varying depths. Channel bottoms usually contain soft mud. Back bays pose major obstacles to vehicular traffic.

Infantry can cross back bays, but only with great effort. If crossing back bays by foot, consult detailed navigation charts and use the following guidelines to plan your route:

- Avoid water less than waist deep; walking in shallow water or soft mud is extremely tiring.
- Avoid back bay islands; these low-lying islands are usually too muddy to support foot traffic.
- Seek out deep water; floating with a pack is less tiring than walking through shallow water or soft mud.
- Seek out sand, shell, gravel, or stone bottoms; these firmer bottoms generally ease travel and help conserve energy.

Rivers and Canals

A river is a large, natural stream of water that empties into a larger body of water. The slope of the riverbed and the volume of water in the river determine its current.

Canals resemble small rivers or streams in their width and depth, but usually lack any significant current. Climbing out of these waterways can be difficult if the canal is flanked by steep banks.

Chapter 5
Fording Waterways

WARNING

Fords are dangerous. Cross as quickly as possible.

A ford is any site in a river, stream, or canal where the water is shallow enough for troops or vehicles to cross without using flotation devices. Canal bottoms are usually too soft to support fording vehicles, and wading infantry frequently stumble.

The tactical situation dictates the location of the fording site. Seek fords that are protected from enemy observation and that allow for adequate supporting fires. A night fording takes at least one and a half times as long as a daylight fording.

> *NOTE:* In this chapter, the term "river" refers to rivers, streams, and canals.

Selection of a Ford Site

The following table identifies desirable fording characteristics:

Characteristics	Comments
Concealment	The ford hides personnel and vehicle movement from enemy observation.
Accessibility	The ford should have low banks with gentle gradients. This allows a free flow of traffic at both the entrance and the exit.

Characteristics (Continued)	Comments (Continued)
Slow Current	The ford's current should not exceed 1.5 meters per second if possible.
Firm Footing	The ford's bottom, entry, and exit composition should be firm enough to support traffic. Do not drive a vehicle over any bottom composition that a 2-inch diameter stick can be pressed into more than 1 or 2 inches.
Gently Sloped Channels	The ford's entry and exit points should be gently sloped. If possible, locate a portion of the stream where the channel is not actively shifting.
Depth	The fording depth is less than or equal to the least capable vehicle.

Determine the Slope

Units move into and out of water faster and more quietly if entry and exit points are not steep or muddy. Slope is the amount of change in ground horizontal distance (run) and in vertical elevation (rise) from one point to another. Slope is usually expressed as a percentage. You can use a clinometer, map, or line of sight and pace to measure the percentage of a slope.

Clinometer

A clinometer measures percentage of a slope. It is a component of the M9 armored combat excavator, and organic to many Marine engineer units. However, engineers are also taught (and often use) field-expedient methods to determine slopes and gradients.

Map

A map measures the horizontal distance along a desired path. Determine the difference in elevation between the path's starting and ending points. From a map's scale, you can determine the distance between two points. From a map's elevation lines, you can determine the difference in height between the same two points. Both figures must be in the same unit of measure (e.g., feet, meter). Divide the elevation (rise) by the distance (run) and multiply by 100.

Rise = 165 - 120 = 45 meters

Run = 200 meters

% Slope = $\frac{45}{200}$ **x 100 - 22.5%**

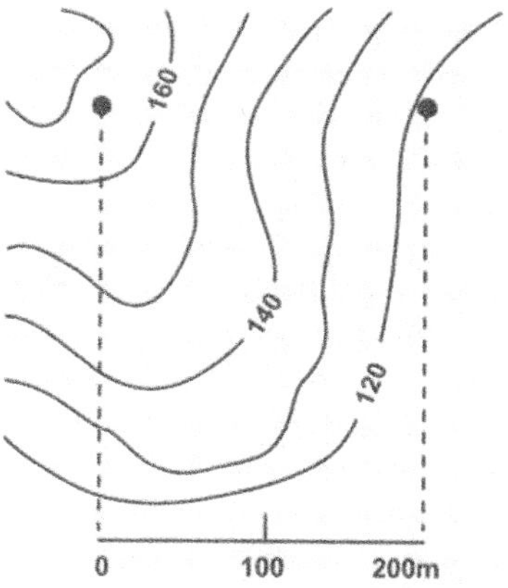

Line of Sight and Pace

To determine line of site and pace, stand at the bottom of the slope, keep your eyes level, pick a spot on the slope, then pace the distance. The number of paces multiplied by a standard measure of 0.75 meter determines the run. The eye-level height (usually 1.5 to 1.7 meters) determines the rise. Repeat this procedure until you have covered the entire distance you want to measure for each spot (vertical and horizontal). Add the vertical distances to provide total rise and the horizontal distances to provide total run.

Determine the Current Speed

Current speed increases as channels narrow. It may be necessary to locate a wider ford location to obtain a slower stream current. The following steps are used to calculate the speed of the current:

Determine points A and B along a channel. Then measure the distance between those two points.

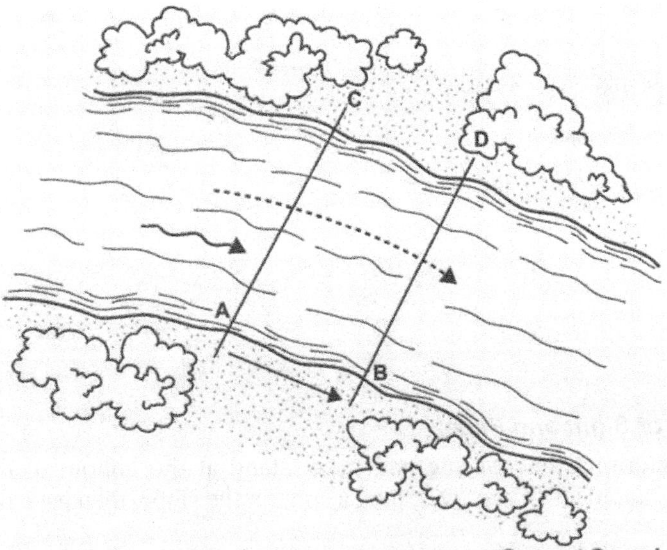

$$\frac{\text{Distance (meters) Between Points A and B}}{\text{Time (seconds) Between Points C and D}} = \text{Current Speed (meters per second)}$$

Sight directly across the water from points A and B to locate points C and D.

Throw a floating object (e.g., a stick) upstream from points A and C. Observe the object as it floats toward points B and D.

Subtract to find the time it takes for the object to float from start to finish.

Do not attempt to swim across currents that are moving faster than 1.5 meters per second. Equivalents of this speed include—

- Quick-time march rate of 120 counts per minute with one, 30-inch step at each count.
- 5 feet per second.
- 3.5 miles per hour.
- 5.5 kilometers per hour.

Measure River Width

A river's width can be estimated from the width of its symbol on a scaled topographic map. If this is not possible, use the following compass techniques:

Stand at the water line (A).

Shoot an azimuth to a point on the opposite bank (B).

Move upstream or downstream until you are at a point (C) where you can shoot an azimuth 45 degrees larger or smaller than the original azimuth.

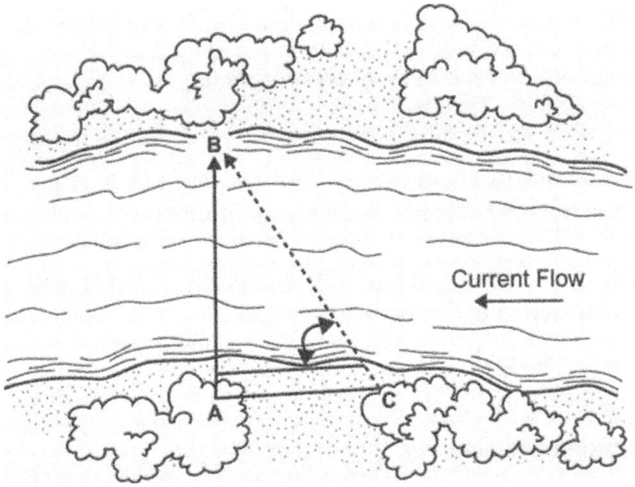

Measure the distance between points A and C. The distance calculated equals the river's width.

Calculate Downstream Drift

A river's current causes personnel and equipment to drift downstream. If personnel and equipment are aimed straight across the river, they will sideslip downstream as they move across the current to the other shore. Therefore, personnel and equipment crossings must compensate for the effects of a river's current and entries are usually made upstream of the desired exit point. Use the formula below to calculate downstream drift.

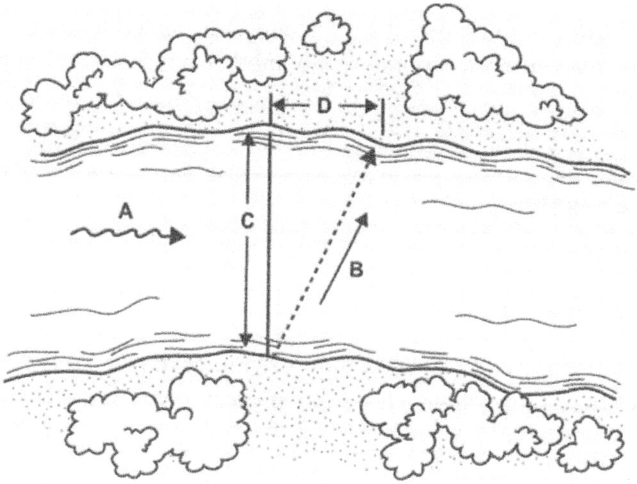

$$\frac{\text{Current Speed (A)}}{\text{Crossing Speed (B)}} \times \text{River Width (C)} = \text{Downstream Drift (D)}$$

> *Note:* The crossing speed for a swimmer across a river may vary but is generally limited to 1 meter per second. All measurements must be in the same unit of measure (e.g., meters, feet).

The Buddy System

Whenever a Marine unit must enter into or operate on the water, a "buddy system" is employed in which every Marine is paired with a swimming partner. The buddy system matches an experienced swimmer with a weak swimmer. The experienced swimmer assists and encourages the weaker swimmer and bolsters confidence during night crossings. If a unit has an odd number of Marines, place the extra person with another pair to form a three-person team.

Water Crossings

Care of Weapons

Marine infantry weapons and munitions are designed to be able to operate after immersion. However, protect your weapons from moisture whenever possible.

A gas-operated weapon can malfunction if water travels down the barrel and enters the gas tube. To protect the gas tube—

Close the weapon's bolt before entering the water.

Seal the muzzle with a condom, balloon, plastic spoon wrapper, or other form of waterproof material.

Tie or melt the protective cover to create a watertight seal.

When the muzzle's protective cover is no longer needed, remove it. Open the bolt and inspect the barrel. If the tactical situation permits, swab excess moisture from the barrel. Test fire automatic weapons, if possible. Field strip and clean weapons as soon as possible. If time does not allow for a complete inspection, rinse inaccessible areas with small amounts of diesel fuel, then dry.

ISOMAT Raft

Construction of an ISOMAT raft is time-consuming. This type of raft should not be employed as part of an attack, but used for logistical purposes (e.g., evacuating stretcher cases, transporting supplies). Use the following steps to build an ISOMAT raft:

Wrap ISOMAT sleeping pads around sturdy sticks.

Use parachute cord and square knots to tie the pads securely in place and to lash stick ends together in a rectangle.

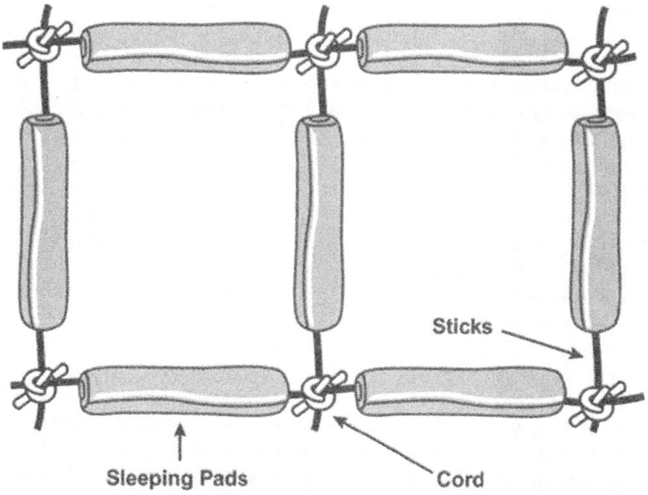

Note: The ISOMAT raft pictured can support several hundred pounds. However, the cargo will get wet if not properly waterproofed.

Poncho Raft

A poncho raft can support two Marines and their equipment and is well suited for long crossings. Use the following steps to build a poncho raft:

Inspect two ponchos and ensure they are serviceable.

Lay one poncho flat on the ground, with the hood-side up.

Cinch the hood tightly to form a gooseneck or tie in a knot.

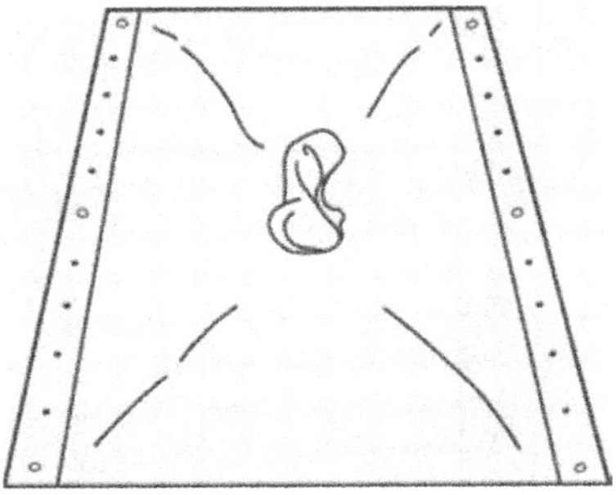

Pad sharp edges of equipment and place the equipment in the center of the poncho.

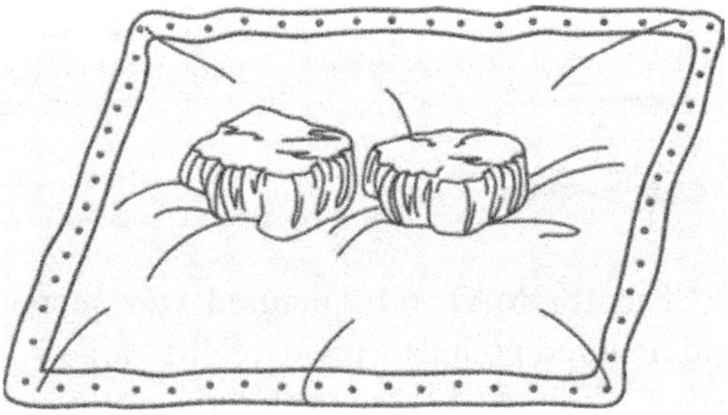

Place the second poncho over the equipment, rubber side up, and hood facing down.

Snap the edges of the two ponchos together.

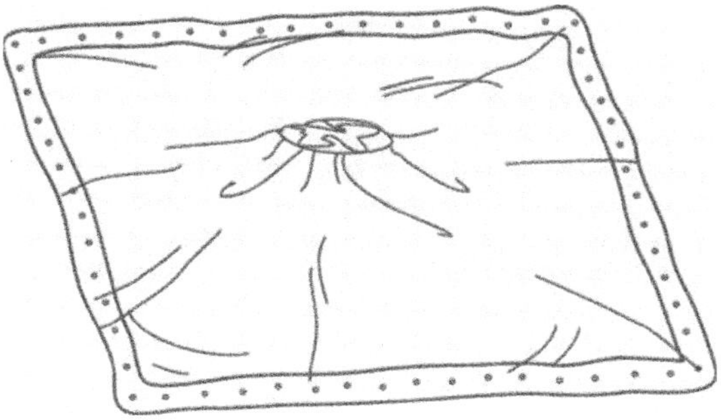

Roll the edges toward the equipment.

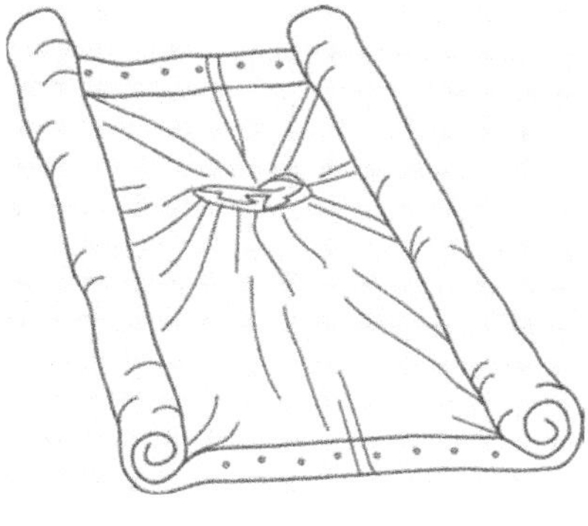

Roll the edges into pigtails and tie them off.

Pull the pigtails together over the top and lash them securely.

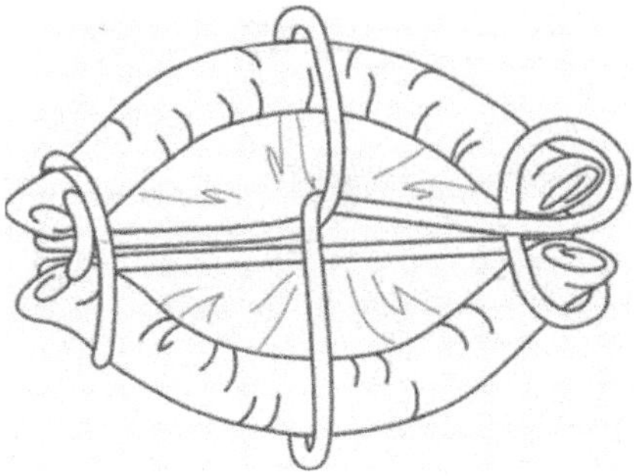

Protect the raft from brush punctures while placing it in the water. Swim across the water obstacle while security elements are covering the far shore.

Construction of a Pack Raft

You will need two waterproofed packs and two M16A2 service rifles. The following steps are required to construct a pack raft:

Place two packs side-by-side with the pack frames on the deck. The tops of the packs are opposite of each other.

Loosen the main compartment straps on both packs.

Insert one rifle on each end between the straps and the packs. The muzzles are opposite of each other. The rifles serve as one means to secure the packs together. Place the front sight post under the top flap.

Tighten the straps so that the rifles and packs are secure.

Take the excess strap on the inner side of each pack and secure it to the opposite pack to better secure the two packs together.

Take the excess straps on the outer sides of the packs and use those straps as safety lashing for the rifles.

Tuck the excess straps and check to make sure the rifles and packs are secure.

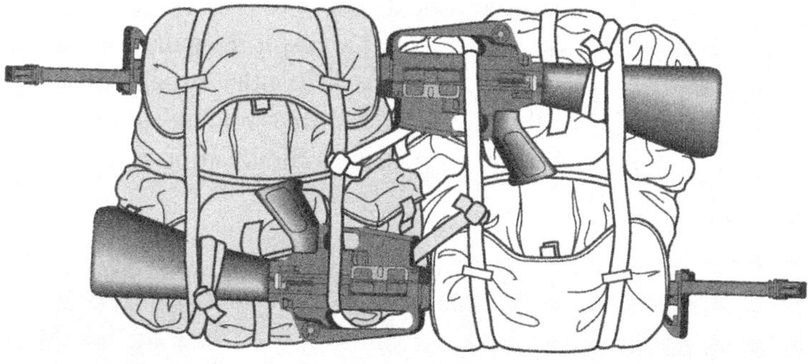

Single-Rope Bridge

A single-rope bridge offers a temporary and quick way to cross small rivers. It also provides extra security while crossing swift waters. At night, it prevents straggling, and guides units precisely from one side of the river to the other side. If crossing a river at night, plan for at least one single-rope bridge.

If your unit is crossing a river with swift currents or water depths above 4 feet, the unit is carrying sufficient rope to span the crossing site, and the tactical situation permits, secure the rope on near and far banks to provide a hand-hold for crossing Marines. This reduces the time required for the entire unit to cross and provides a degree of comfort/confidence for poor swimmers. Use a squad-sized bridge team to construct a single-rope bridge. Station several strong swimmers at the water's edge to help anyone who has trouble crossing.

Nylon rope is normally coiled in 120 foot lengths. It is 0.6 inches in diameter and has a breaking strength of about 3,840 pounds. Over time, a nylon rope can stretch to as much as one-third more than its original length and stretching weakens the rope. If the rope is stretched, discard the rope or use it for light tasks. To prolong the life of a nylon rope, do not step on it or drag it on the ground. Pad the rope in places where it contacts rocks or sharp corners. Do not leave the rope knotted or stretched longer than necessary. Dry rope as soon as possible. Single-rope bridge construction is as follows:

Tie a sling rope around your waist using a square knot and two, separate half hitches. See the appendix for detailed information on knots.

Attach a locking steel carabiner to the sling rope.

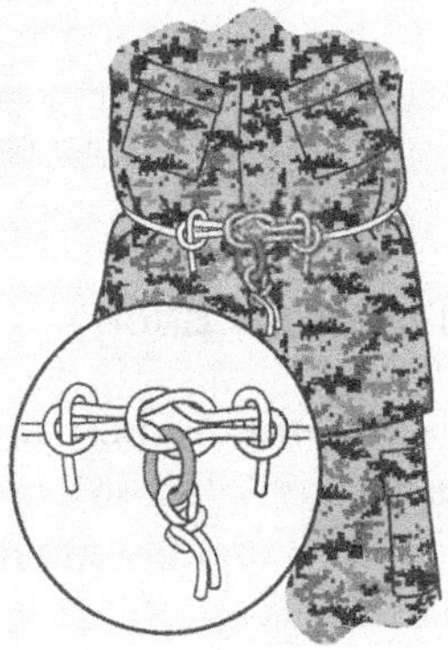

Tie a bowline knot in the running end of the bridge rope and attach it to the carabiner.

Temporarily secure the other end of the rope to a tree on the near shore.

Enter and cross the water.

NOTE: Carry only your weapon and ammunition.

Exit the water on the opposite shore.

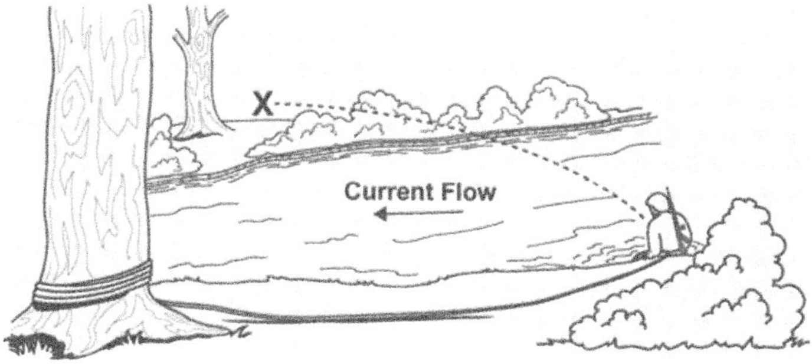

Prepare your weapon for use. Unhook the bridge rope from the carabiner at your waist, and tie the bridge rope to a sturdy tree using a round turn and two half hitches.

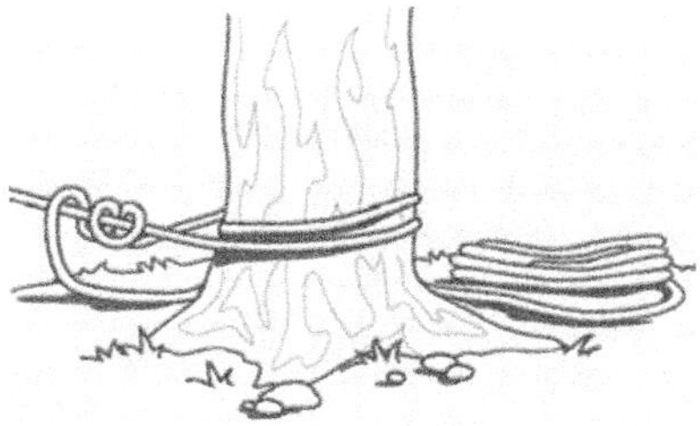

Conduct a box reconnaissance of the opposite shore.

On the near shore, have another Marine prepare to tighten the rope. That Marine should place a transport tightening system in the bridge rope by tying a double butterfly knot and placing two carabiners in the butterfly. See the appendix for detailed information on knots.

The Marine should pass the running end of the bridge rope around the downstream side of the near shore anchor point and through the two carabiners.

Pull the butterfly knot approximately one-third of the distance across the river.

Secure the bridge rope to an anchor point using a round turn and two half-hitches.

On the near shore, the Marine helping you should pull the slack out of the bridge rope until the butterfly knot is back on the near side. The bridge rope is then tied off against itself using two half hitches with a quick release in the last half hitch.

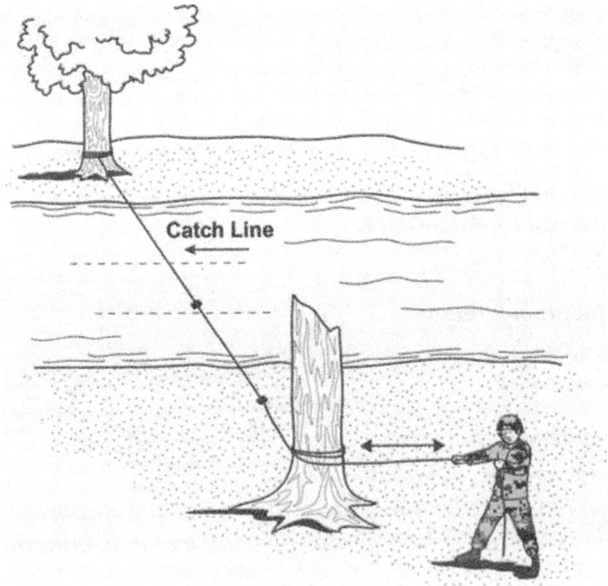

NOTE: The single-rope bridge must be as tight as possible so it will not sag when used.

If you lose your footing and fall into the water, swim with the current to the closest shore. Swimming against the current is dangerous and quickly causes fatigue.

High and Dry Crossings

If the single-rope bridge is high enough, suspend yourself below the single-rope bridge and above the water. Use the following steps to suspend yourself from a single-rope bridge and then pull yourself across the water:

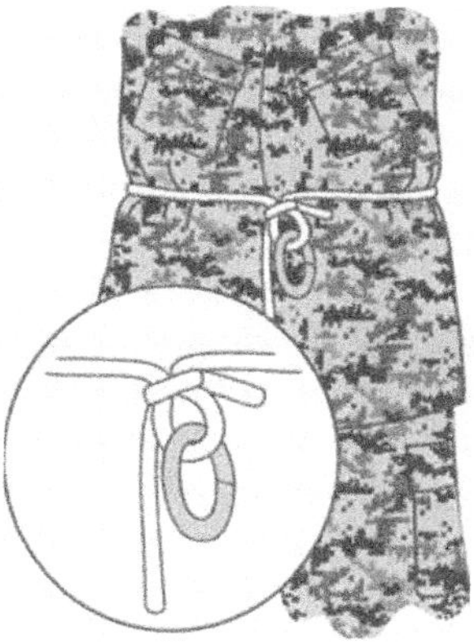

Tie a sling rope around your waist using a bowline. Ensure that the knot is tight.

Attach a carabiner through the bowline's loop. The carabiner's gate faces up.

Secure your helmet chin strap.

Face the single-rope bridge with your left shoulder toward the far shore.

Grasp the bridge rope in both hands.

Swing your body beneath the single-rope bridge with your head toward the far shore. Cross your ankles above the bridge rope.

Arch back until the carabiner contacts the bridge rope. Connect the carabiner to the bridge rope. Allow the carabiner to bear your body's weight.

Pull yourself across the single-rope bridge, hand over hand, to the far shore.

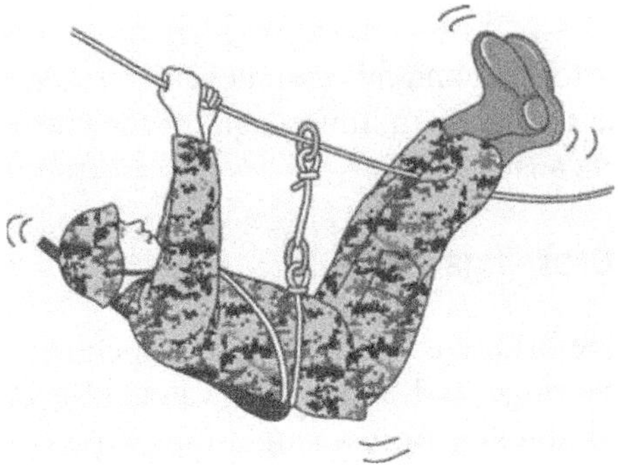

Swift Current Crossings

A single-rope bridge prevents being knocked down and swept away by a swift current. Use the following steps to move through a swift current:

Tie one end of a sling rope around your waist using a bowline.

Tie the running end of the sling rope in another bowline, and attach a carabiner to the bowline's loop.

Step up to the bridge. Face upstream.

Hook the carabiner to the single-rope bridge.

Walk sideways into the river while grasping the bridge rope in both hands.

Use the single-rope bridge for balance and remain standing, if possible.

Continue to move sideways through the river to the far shore.

Slow Current Crossings

If you face little or no current, it is not necessary to hook up to a bridge rope with a carabiner. Lie on your back in the water beneath the single-rope bridge. Support your body weight with your waterproof pack. Use the bridge rope and pull yourself hand over hand across the river.

Removal

If you are the last Marine waiting to cross, pull on the standing end of the rope to release the knot, then tie the rope around your waist using a bowline. The Marines on the far shore will pull you through the water.

Appendix
Knot Tying

Square Knot

The square knot is used to secure two ropes of equal diameter together so they form one continuous rope that will not slip.

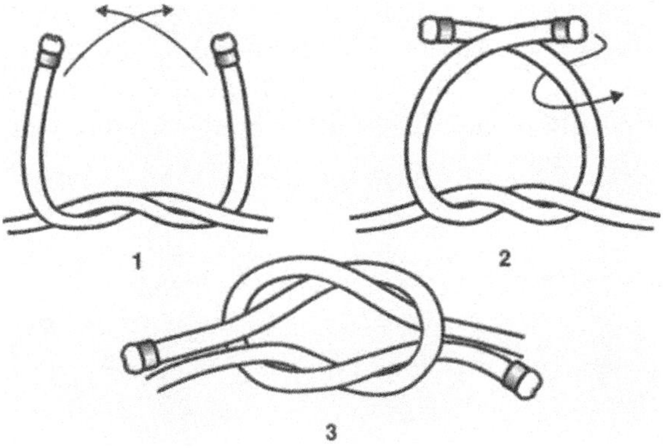

Bowline

The bowline forms a loop that will not tighten or slip under strain. It is easily untied.

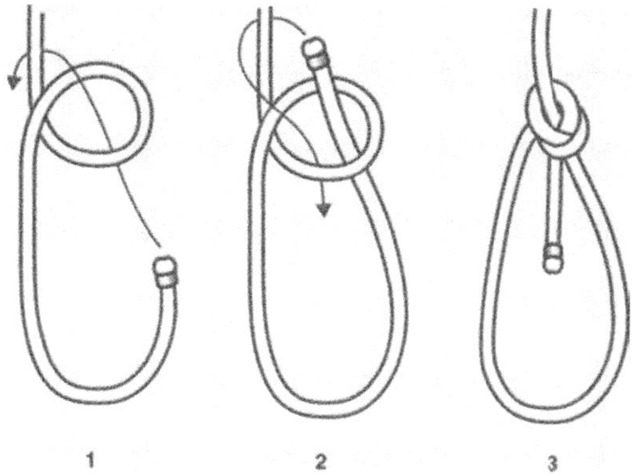

Hitches

Hitches are used to dress (prepare) knots and secure loose ends.

Half Hitch

The half hitch is used to tie a rope to a tree or to a larger rope. It will hold against a steady pull, but is not a secure hitch. It is frequently used to secure the free end of a rope.

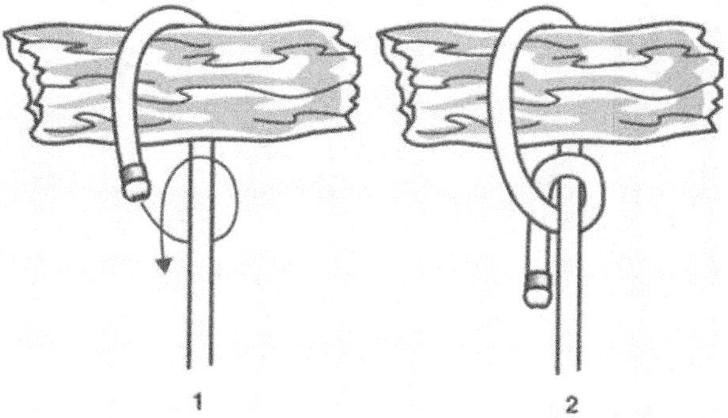

Two Half Hitches

Two half hitches can be used to secure the running end of a rope.

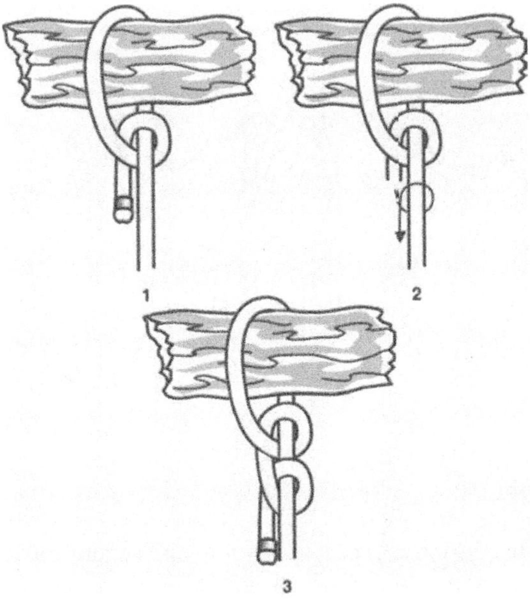

Round Turn and Two Half Hitches

A round turn and two half hitches can be used to fasten a rope to a tree. This hitch does not jam.

Butterfly Knot

The butterfly knot is used for anchor lines. Tied properly, it will not tighten on itself to the point that it cannot be easily untied. The butterfly knot can be used to pull a rope bridge taut. This knot can be used to tighten a fixed rope when mechanical means are not available. It will not jam if a stick is placed between the two upper loops.

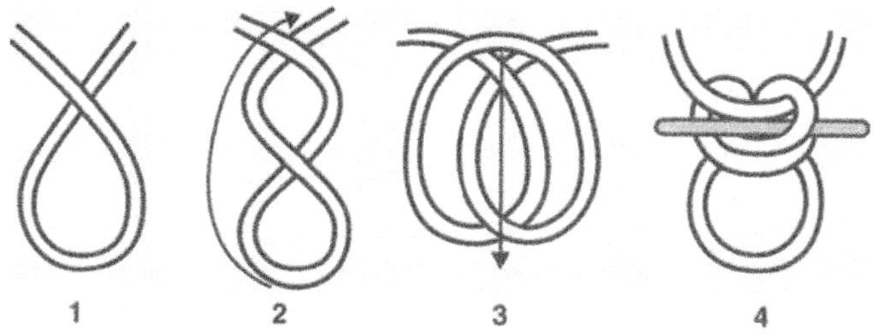

TRAINING CIRCULAR
No. 21-3

HEADQUARTERS
DEPARTMENT OF THE ARMY
Washington, DC, 17 March 1986

SOLDIER'S HANDBOOK FOR INDIVIDUAL OPERATIONS AND SURVIVAL IN COLD-WEATHER AREAS

TABLE OF CONTENTS

*** This publication supersedes TC 21-3, 30 September 1974.**

CHAPTER 4. SMALL ARMS AND AMMUNITION

CHAPTER 5. RATIONS AND DIET

CHAPTER 6. HYGIENE AND FIRST AID

CHAPTER 7. BIVOUAC ROUTINE

CHAPTER 8. INDIVIDUAL MOVEMENT

CHAPTER 9. LAND NAVIGATION

CHAPTER 10. SURVIVAL

PREFACE

This handbook is an aid for soldiers in cold climates. However, expertise in dealing with cold-weather operations cannot be gained just from reading this handbook. To develop successful cold-weather field techniques, application training and operational experience are necessary. Additional information is available in FM 31-70, FM 31-71, FM 90-11 (TBP), and FM 9-207.

Military history of operations in northern areas shows that man's success or failure is measured by his regard for the dominant characteristics of a cold-weather environment. For a soldier to become a cold-weather expert, he must learn the techniques and rules needed to fight and live in a cold-weather climate.

The proponent of this publication is HQ TRADOC. Submit changes for improving this publication on DA Form 2028 (Recommended Changes to Publications and Blank Forms) and forward it to Commandant, US Army Infantry School, ATTN: ATSH-B, Fort Benning, Georgia 31905-5593.

Unless otherwise stated, whenever the masculine gender is used, both men and women are included.

CHAPTER 1

Introduction to Cold-Weather Operations

> ***"Soldiers in pairs look after each other; be faithful to your mate; be loyal to your leaders."***
>
> ***Anonymous***

1-1. COLD CONDITIONS

COLD-WET

Cold-wet conditions occur when temperatures are near freezing and high and low temperatures cause alternate freezing and thawing. These conditions can be accompanied by wet snow and rain, causing the ground to become slushy and muddy. Soldiers need clothing that has a waterproof or water-repellent, wind-resistant outer layer, and an inner layer with insulation that provide protection in moderately cold weather (above 14°F). **Waterproof footwear is essential.**

COLD-DRY

Cold-dry conditions occur when average temperatures are lower than 14°F. The ground is usually frozen and the snow dry. Low temperatures plus wind increase the need for protection of the entire body. For these conditions, soldiers need clothing that will provide insulation for a wind-chill factor of -80°F. For temperatures below 14°F, white vapor barrier (VB) boots effective down to temperatures of -50°F must be worn.

WIND CHILL

Wind chill is a measure of the combined effects of wind and temperature. Temperature alone does not give a true indication of the impact of the cold. In order to effectively gauge the difference between temperature and the impact of the cold, a wind-chill scale must be used. The chart in Appendix A is a simple and practical guide showing the conditions under which cold weather is dangerous and when exposed flesh is likely to freeze. Soldiers must be familiar with the wind-chill chart to recognize these dangers.

1-2. EFFECTS OF COLD WEATHER ON MILITARY OPERATIONS

Cold has a numbing effect on skin, which hinders task performance. This must be considered when accomplishing such routine tasks as vehicle maintenance, and making and striking bivouac areas (FM 9-207).

When conditions become extreme and the temperature drops, the problem of survival becomes critical. Under these circumstances soldiers may tend to withdraw emotionally, not wanting to leave the warmth and protection of their shelter or tent. The remedy is physical activity. Try to remain alert and active. Execution of all military duties and the fast and proper performance of group tasks are vital.

COCOON-LIKE EXISTENCE

Many soldiers, when wearing several layers of clothing and with their heads covered, tend to withdraw within themselves and assume a cocoon-like existence. When so clothed, many soldiers' hearing and field of vision are restricted; they tend not to notice their surroundings. Their thinking and reasoning become sluggish.

INDIVIDUAL AND GROUP HIBERNATION

This process is characterized by individuals seeking the comfort of sleeping bags, and by the group remaining in tents or other shelter. Guard and security measures may be neglected, jeopardizing the security and safety of the unit.

1-3. OVERCOMING THE COLD

The body must be protected. It must be kept clean, dry, and warm, with normal body processes maintained. Rest and nourishment are vital. The right approach to cold-weather living will keep you healthy and get you through challenging times. There are four basic rules to follow:

KEEP IN SHAPE

Cold-weather clothing and equipment are heavy, and add weight to your normal equipment load. This hinders movement in snow and uses too much energy. The importance of being in excellent physical condition cannot be overemphasized. Your unit must have a tough, challenging, demanding physical training program.

DRINK PLENTY OF WATER

Water may be hard to get, so you may drink less than you need or only when you are thirsty. This will not give you the water needed to avoid dehydration. Drink plenty of water (at least 3.5 quarts per day) to avoid dehydration and fatigue. DO NOT eat snow as a water substitute; it will lower your body's core temperature.

EAT TO KEEP FIT

Regular, nutritious hot food is needed for top performance. You must eat, even when you are not hungry.

KEEP A GOOD ATTITUDE

You will meet many new challenges, but none that cannot be overcome. Your attitude toward the cold will reflect those of your leaders.

1-4. POSITIVE LEADERSHIP AND THE RIGHT ATTITUDE

Most soldiers, after 3 to 4 weeks of cold-weather training, will be able to cope with the environment. Some soldiers may not be as skilled as others, but most will succeed.

Set an example for other soldiers. Winter conditions can be frightening, trying to meet challenges you have never met before. Any mistake could make you a casualty. Aggressive leadership actions that overcome the challenges of the environment are essential to mission accomplishment.

There are two enemies to overcome in northern operations: the enemy soldier and the cold. The first step in defeating either enemy is having the right attitude. The cold can defeat you psychologically if you are not aware of the symptoms of a poorly prepared unit. The following are suggestions on how to defeat the cold:

- If you become depressed or moody, or do not want to talk, force yourself to get involved with others. Do not leave out the weaker soldier.
- If you find it hard to remember things you have been taught, review orders and drills.
- Remind soldiers that their jobs are to close with, engage, and gut the enemy, and that weapons and equipment must be kept in fighting order. Operations in a northern environment can easily become a camping trip with serious consequences. You must avoid this common ambush.
- Cold weather may slow down the following of an order, but it is not an excuse for not doing it.
- Cold weather may make tasks harder and take longer to do, but it will not make them impossible. With knowledge, proper equipment, and proper training you can overcome the cold and be successful in combat.

CHAPTER 2

Personal Clothing and Equipment

"The freezing German troops were reduced to removing clothes from enemy corpses, improvising strawboots, and taking other emergency measures."

Russia, 1942

2-1. INDIVIDUAL CLOTHING

Soldiers should understand the design principles of the military cold-weather clothing system. These principles are: insulate, layer, and ventilate. There are four ways to make these principles work for you: keep it Clean, avoid Overheating, wear it Loose, and keep it Dry. By remembering the "buzz word" COLD, you can quickly check your clothing.

INSULATE

Insulation material reduces the amount of heat loss. By regulating the amount of insulation, an individual can regulate the amount of heat lost or retained.

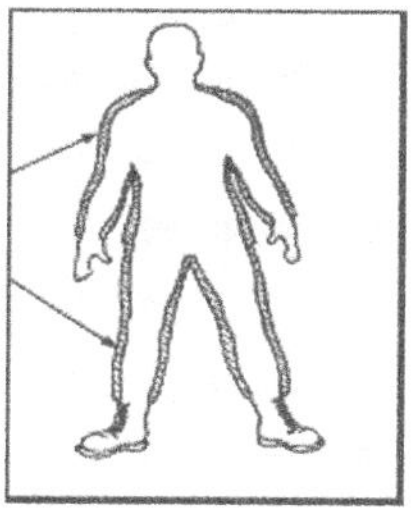

Figure 2-1. Insulation.

LAYER

Several layers of clothing provide more insulation and movement than one heavy garment, even if the heavy garment is as thick as the combined layers. Air pockets trapped between the layers provide insulation and warmth.

VENTILATE

Ventilation helps maintain a comfortable body temperature by allowing excess heat and moisture to escape.

Figure 2-2. C-0-L-D.

KEEP CLOTHING CLEAN

Dirty clothes are cold clothes. Dirt and grease clog the air spaces in clothing and reduce insulation.

AVOID OVERHEATING

Select comfortable clothing and do not overdress. In cold weather, it is better to be a little cool rather than risk being overheated.

WEAR IT LOOSE

All items of the cold-weather uniform are large enough to allow wearing of the appropriate number of layers. Your field jacket may appear too large without all of the layers designed to fit under it. If the uniform items do not fit loosely, the insulation that keeps you warm will be greatly reduced.

KEEP IT DRY

It is important to keep clothing dry, since wet clothing conducts heat away from the body more quickly than dry clothing. Moisture soaks into clothing from two directions: the inside and the outside.

From the outside you must guard against melting snow; from the inside you must guard against sweating. Brush snow and frost from clothing before entering any heated shelter or vehicle, and avoid overheating.

2-2. COLD-WET VERSUS COLD-DRY

As mentioned in Chapter 1, temperatures that are above 14°F (-10° C) are considered cold-wet conditions; temperatures consistently below 14°F will cause cold-dry conditions. Cold-wet conditions usually involve a freeze/thaw cycle, often accompanied by rain or wet snow. This type of weather is hard to dress for, since each soldier must battle the elements to keep his clothing dry. During cold conditions, the ground is usually frozen and the snow is dry. Athough the temperatures during the cold-dry conditions may be lower than during cold-wet conditions, it is easier to stay warm.

COLD-WET		COLD-DRY
	OR	
DOWN TO 14°F		BELOW 14°F

Figure 2-3. Cold-wet versus cold-dry.

2-3. THE COLD-WEATHER UNIFORM

There are many layers of clothing that must be worn in a cold weather environment. The first layer of clothing is the T-shirt and shorts. The second layer is a set of thermal underwear, which consists of an undershirt and underpants. You should wear only the clothing needed to keep comfortable and warm.

The scissors-type suspenders must be the last thing next to the outer layer, or else the layers cannot be tucked inside the trousers. The underpants and all succeeding layers of trousers are supported from the metal hooks on the suspenders. The suspenders allow the underpants and trousers to be worn loosely at the waist so that circulation and ventilation are not restricted.

Wear the wool shirt outside the trousers, which provides another method of ventilation. You can increase the waist size of this shirt by removing the stitching in each dart at the sides of the shirt.

Wear the field trousers as an outer garment in the cold-wet environment. In the cold-dry environment, wear the nylon liners inside the field trousers during periods of inactivity. On the inside of each cargo pocket on the trousers is a tape that runs through a hole on the side of the pocket. The tape ties around the thighs and keeps the cargo in the pocket from rubbing and irritating the legs. For warmth and to prevent entry of snow, tie the drawstrings around your ankles outside the boots into the boot lacing, and tuck the bowknots under the trousers.

The field jacket is worn with the liner in both cold-wet and cold-dry areas. Wear the coat and liner directly over the wool shirt. Remove the liner in mild weather or when activity increases.

In a cold-dry environment the parka is worn with a liner. The waist is adjustable for ventilation and warmth with a drawcord, and the skirt has a "split-tail" that can be tied tightly around each leg for warmth. During a period of severe cold, when activity is less, wear the parka with liner over your field jacket with liner. Remember: do not overdress or you will sweat.

Do not dry nylon on drying lines too close to the heat; too much heat will melt the nylon, destroying its insulating value.

Keep the head covered to stay warm; a warm head even warms the feet. Wear the cold-weather cap or balaclava under your helmet or under the hood to your field jacket. When wearing the cold-weather cap as an outer headpiece, the lower flap may be folded up over the top and fastened with hook-pile tape straps in front. In a cold-dry environment, wear the winter hood over the cap or the helmet. Adjust the hood by using the drawstrings on either side of the hood and the elastic strip at the back. To provide minimum protection for the face from the wind, bend the wire in the fur cuff of the hood so that you have a small opening for your face. This allows visibility but less protection.

The feet are hard to keep warm and dry, so protect them. The insulated boots (black for cold-wet conditions and white for cold-dry conditions) are warm and waterproof. The white insulated boot is the best cold-weather footwear today. When you are active, your feet will sweat and, because of the waterproof rubber lining, the moisture will remain in the socks or in the bottom of the boots. The feet, even if damp, will stay warm because the heat given off by your body will remain inside the boots. If the boots are worn for long periods, the feet become white and wrinkled when you remove the boots. This is no cause for alarm. Drying, massaging with foot powder, warming and putting on dry socks will correct this. If you are inactive for long periods, your feet may become cold. Exercises such as knee bends, stamping your feet, running in place, or wiggling your toes will make them warm. Elevate the feet whenever possible to aid blood circulation. Don't wear socks that are too tight, and wear only one pair of cushion-sole socks at a time. Change your socks at least twice a day. Most important, stay active.

Care of your boots is important. Wash boots inside and out with soapy water at least once per month or as required by unit SOP you want your feet to keep warm, you must keep the insulation dry. Inspect your shoes often and repair any holes or punctures immediately. Temporary repairs can be made by patching the hole with any kind of tape; even chewing gum can be used. Seal the hole quickly to prevent the insulation from becoming wet. If one of your boots feels heavier than the other, the insulation is probably wet and they should be turned in for new ones.

Insulated boots must be tested annually to ensure they are waterproof and have less than 0.01 percent moisture in the insulation. The air valve on the outside of the boot is used to equalize pressure when worn at high altitudes and adjusted again upon return to low altitudes. To equalize the pressure in the boot, open the valve for a few seconds and immediately close it. The valve should be kept closed at all other times to prevent moisture from entering. Never open the valve unless the boot pressure around your foot causes discomfort.

You can wear overshoes over the leather boots on wet or muddy terrain. The overshoes are not as warm and dry as the insulated boots. Make sure your socks are dry and that your boots are loosely laced so as not to restrict circulation.

2-4. EXTENDED COLD-WEATHER CLOTHING SYSTEM (ECWCS) (TEST)

The ECWCS is issued as a one-for-one replacement for the current standard cold-wet clothing system. System components use the most recent materials and technology. The result is a reduction in weight and bulk from the current cold-wet clothing system without loss of protection from the cold. The ECWCS uses the layer approach that draws moisture away from the body and should be worn as discussed below.

FIRST LAYER

This consists of polypropylene undershirt and shorts. They are worn next to the skin. They are designed to draw moisture away from the skin, thereby reducing conductive heat loss. You must make sure that no other item is placed between these items and your skin.

SECOND LAYER

The second layer contains a synthetic fiber pile (SFP) shirt and polyester batting trouser liners. The SFP shirt and trouser liners provide insulation. The shirt is designed to dry quickly if it becomes wet and to provide a high degree of insulation. It should not be worn as an exterior garment in wet or windy conditions, since the loose weave provides little protection against wind chill. If the shirt gets wet, it will reduce the insulation value.

The trouser liners are designed to be worn during little or no activity. They should not be worn during heavy activity such as skiing or snowshoeing.

THIRD LAYER

This consists of polyester batting field jacket liner and field trousers. The jacket liner should only be used for extra insulation during periods of little activity. This liner is designed to be used as a separate jacket; it should not be attached to the parka. The trousers provide a durable wind-resistant, water-repellent layer to protect the insulating underpants and trouser liners.

FOURTH LAYER

Windproof, waterproof parka and trousers make up this layer. The parka and trousers provide windproof and waterproof shell garments to protect the first three layers.

The parka should be the outer garment, which is always worn when doing any activity outside a shelter. It is waterproof, yet the fabric is breathable so that body moisture can escape. The hood is designed to give full protection from wind and snow and does not prevent your seeing the area around you. It has underarm zippers to allow air to circulate during heavy activity. The trousers are made of the same material as the parka and are very warm. They should be worn only when necessary to prevent the field trousers from becoming wet.

FIFTH LAYER

This consists of standard camouflage overwhites.

2-5. TIPS ON WEAR AND MAINTENANCE OF CLOTHING

The previous discussion on the layers of cold-wet and cold-dry uniforms describes how each system functions. In some units, the cold-dry field trouser liners are often substituted for the wool field trousers of the cold-wet system and function well in a cold-wet environment. Your commander and unit standing operating procedure (SOP) will determine what uniform and components will be worn and carried in the field.

DRYING WET CLOTHING

Hang items separately from drying lines inside the tent. Do not place anything too close to a stove or over steaming pots; steam will cause further dampness.

DRYING DAMP CLOTHING

Damp items can be hung on the rucksack during movement. Pin damp socks to load-carrying equipment or shirt during movement or just inside the last layer of clothing. Do not place damp clothing in sleeping bags. This will only make sleeping bags wet, and they are harder to dry than clothing.

Small items can be placed between the sleeping bag and the sleeping mat under your legs. When washing clothes without soap, dry-rub and air clothing and sleeping bags.

Never warm your feet next to an open fire. The boots will melt and burn because of their insulation before the feet become warm.

Try never to wear VB boots in a sleeping bag. Sometimes the tactical situation is such that boots must be worn while sleeping. If so, the feet must be dried and massaged with foot powder, and dry socks must be put on. Wearing the VB boot while sleeping does not let your feet dry out. After VB boots have dried, place them underneath your sleeping bag on top of your sleep pad when temperatures are below 0°F.

2-6. COLD-WEATHER EQUIPMENT

The sleeping equipment consists of a sleeping bag, an insulated sleeping pad, and a waterproof bag. There are two types of sleeping bags:

(1) Sleeping bag type I, intermediate cold, temperature range +45°F to +10°F, consists of a mummy-shaped bag with hood, insulated sleeping pad, and a waterproof bag.

(2) Sleeping bag type II, extreme cold, temperature range of +10°F to -50°F, is also available.

Whatever clothing is necessary must be worn with both bags to stay warm and comfortable at the lower temperature ranges. As a minimum, long underwear, sleep shirt, dry socks, and head protection should be worn while sleeping.

The insulated sleeping pad has replaced the pneumatic mattress. Use the insulated sleeping pad under the sleeping bag for warmth and comfort and to keep your sleeping bag dry.

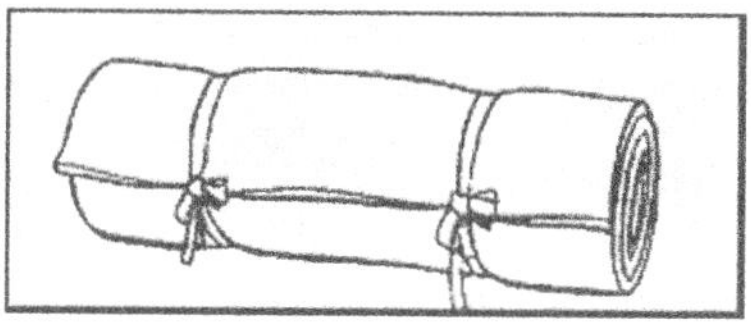

Figure 2-4. Sleeping pad.

The bag must be kept dry if it is to keep you warm. Sleeping bags become wet from outside moisture, from sweating on the inside of the bag, and from water escaping from the mouth or nose into the sleeping bag.

To keep outside moisture from wetting the bag, place protective insulation, pine or spruce boughs, a poncho, or an insulated sleeping pad under the bag. Avoid sweating by wearing the least amount of clothing to keep warm inside the bag, and by using the proper sleeping bag for the temperature conditions. Do not breathe into the bag or moisture will collect and wet the bag or form ice crystals. If your face gets cold, cover it with a towel, muffler, balaclava, or sleeping hood. The sleeping shirt with hood is specifically designed for use with the types I and II sleeping bags.

When it is not in use or being carried, open the sleeping bag wide so that fresh air can get into it. Turn it inside out so that the sun and wind can dry it. Always shake the bag out and fluff it up before and after use to provide for more trapped air between the layers of the wind-resistant/water-resistant outer shell.

Always carry your sleeping bag in the waterproof carrying bag. This protects the bag and helps keep it dry. Ensure that the top bag is securely tied to prevent snow or other moisture from getting inside and wetting the sleeping bag. The following suggestions will help you keep warm and comfortable in the sleeping bag:

- Eat just before getting into the bag. This gives you energy and helps you to stay warm.
- Always relieve yourself just before going to bed. The act of getting out of a warm sleeping bag to go outside can be a chilling experience.

PACKING YOUR RUCKSACK

When you load your rucksack for skiing or climbing, pack heavy items at the bottom and next to the frame. This places the weight on your hips, which is necessary for good balance. Place hard or sharp objects inside the load where they will not rub on the bag and your back. When walking or snowshoeing, raise the center of gravity and take more of the load on your shoulders. Articles that are often needed are placed in the outside pockets for easy reach. Keep maps and other flat objects in the flap pocket.

ADJUSTMENT

Adjust the straps so that you can move the top part of your body and swing your arms freely. Adjust the shoulder straps so that the lower back straps fit just above the belt line.

2-7. LOAD-CARRYING EQUIPMENT

The all-purpose lightweight individual carrying equipment (ALICE) rucksack comes in sizes medium and large. Make sure you are equipped with the large ALICE in a cold environment.

In cold-weather operations, you should follow the packing list set forth in your unit SOP to organize your rucksack. The main consideration is to avoid having to unpack during halts to get frequently needed items. The following illustrates a method of organizing your rucksack:

- **Small external pockets.** These contain small high energy foods (candy, cereal bars) to be eaten while on the move.
- **Large external pockets.** These contain rations to be eaten for morning and evening meals. Take extra socks, scarves, and a spare cap. The openings behind the pockets can be used for skis or other items.
- **External attachment points.** The sleeping pad is attached to the bottom of the pack or under the top flap (wrapped inside sandbags) and other combat items as required.
- **Top flap.** It contains camouflage overwhites and pack cover.
- **Main compartment.** Place the sleeping bag in the bottom of the rucksack. Spare clothes are placed in the upper half of the compartment for easy reach.
- **Camouflage.** A white cotton cover is provided that covers the entire rucksack, excluding the shoulder straps. Use it when wearing your overwhites.
- **Care and maintenance.** Check your rucksack for damages before and after an operation. Repair damages immediately or turn in through supply channels for a replacement.
- **Emergency quick-release of the rucksack.** When quick removal of the rucksack is needed, unsnap the fastener. Pull down on the quick-release fastener on the left shoulder strap, shift the rucksack toward the right shoulder, and let it drop to the ground.

- **Cargo support shelf.** The rucksack can be converted to a cargo carrier by removing the combat pack and replacing it with the cargo support shelf. You can carry square, rectangular, or box-type items such as radios, water cans, and fuel cans.

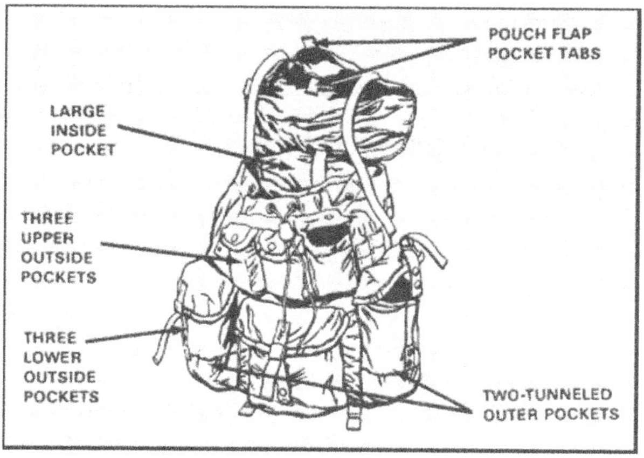

Figure 2-5. Load-carrying equipment.

2-8. OVER-THE-SNOW MOVEMENT EQUIPMENT

Before you are instructed to use snowshoes or skis, your chain command will first determine if movement by foot is possible. In shallow snow (8 inches or less) or heavily crusted snow, movement on foot for inexperienced skiers provides better maneuverability with the least effort.

SNOWSHOES

Snowshoes consist of a white magnesium frame laced with nylon-coated steel cable. They are about 48 inches by 12 inches weigh 3.3 pounds with bindings. When properly adjusted, the toe of the boot will be about halfway across the opening in the snowshoe; the toe strap should be firm, but not tight enough to restrict circulation; and the heel straps should permit some side-to-side movement. When used correctly, snowshoes provide better mobility than walking. In most instances, skis are preferred over snowshoes. When using snowshoes, you cannot back up or swim; be careful when moving over thin ice. The traction needed to ascend or descend with snowshoes can be increased by waxing. The use of ski poles with snowshoes aids movement and reduces fatigue.

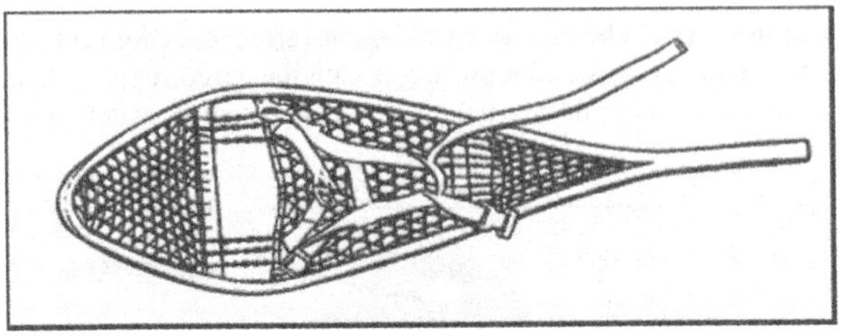

Figure 2-6. Snowshoe and snowshoe binding.

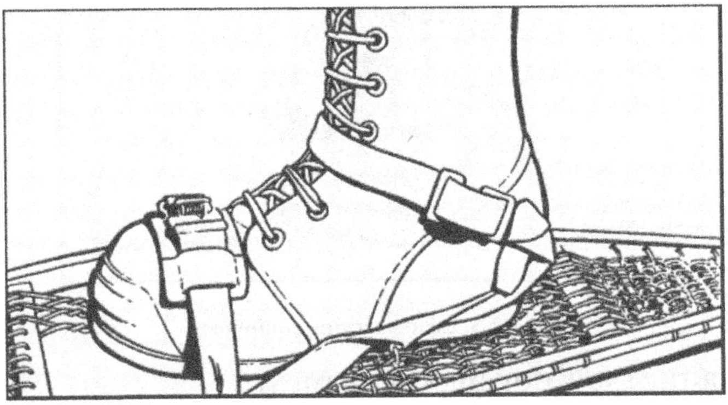

Figure 2-7. Snowshoe binding attached to VB boot.

SKIS

The standard military skis are about 208 centimeters (82 inches) in length, and have a hole in the tip for towing, a groove in the heel for attaching mohair climbers, and metal edges.

SKI POLES

Ski poles can be very useful in steep terrain. The military ski pole consists of a steel shaft with rubber handgrips and baskets, and a leather wrist strap. The poles come in three lengths: 130 centimeters (51 inches), 137 centimeters (54 inches), and 147 centimeters (58 inches). When fitting the poles, you should stand on the floor and place the pole at your side; it should come to the point of your shoulder (about 1 inch below the top of the shoulder).

MOHAIR SKI CLIMBERS

The mohair climbers are attached to the bottom of the skis for added grip when climbing steep slopes. The climbers consist of a cotton-backed mohair pile and are attached to the skis by five web straps.

2-9. MISCELLANEOUS EQUIPMENT

SUNGLASSES

Polarized sunglasses should always be worn on bright days when the ground is covered by snow. Failure to wear the sunglasses, even on overcast days when the need seems less, may result in blindness.

PROTECTIVE MASK

There is an M4 cold-weather winterization kit available for your protective mask. Make sure you install this winterization kit for operations in the northern environment.

CANTEENS

In cold weather the plastic canteen will freeze very quickly if carried in the standard fabric carrier. Since the canteen cannot be placed directly on heat, there is no quick method for thawing. Whenever possible, the canteen should be carried in one of the interior uniform pockets, or wrapped in clothing and placed in the rucksack to ride against your back. In extreme cold, do not fill the canteen over two-thirds full; this allows for expansion if the water freezes.

The insulated canteen is a one-quart capacity stainless steel vacuum bottle. It is provided with a cotton duck cover and a stainless steel cup. A new arctic canteen will keep water from freezing at -40°F for 6 hours. The following are some tips in using the arctic canteen:

- Fill the canteen using your canteen cup.
- Fill to within half an inch of the neck.
- Do not drop or dent.
- Put it in your sleeping bag at night.
- Carry the canteen in an M60 bandoleer near your body.

Figure 2-8. Insulated canteen with cup.

CHAPTER 3

Tents and Heating Equipment

"A well-sheltered defender enjoyed a marked advantage over his exposed attacker..."

Ironside, 1918

3-1. GENERAL

To conduct successful military operations in cold weather and maintain a high level of combat efficiency, you must have heated shelters. Tents and stoves are a vital part of cold-weather equipment.

3-2. TENT GROUP EQUIPMENT

The following is a list of the tent or ahkio group for an infantry squad:

1 Ten-man tent with liner
1 Yukon stove
1 Metal or plywood stove base
2 Five-gallon gasoline cans (one with white gasoline per platoon)
1 Five-gallon water can (filled)
2 Cases, meals, ready-to-eat
1 Climbing rope (120 feet)
1 Box of candles
2 Squad cook sets
2 One-burner squad stoves
2 Axes
2 Saws
1 Hatchet with sheath
1 D-handle shovel
1 Sledgehammer
1 Pick
1 Tripod for gas can (set)
3 Pairs of snowshoe bindings (spare)
1 Lantern (extra mantles)
1 Box of soda bicarbonate (fire extinguisher)
1 Roll of soft wire
5 Plastic bags
3 Rolls of toilet paper

3-3. AHKIO

The ahkio is the squad's method of transporting the tent group equipment in the winter. It is a 200-pound capacity fiberglass sled that can be used for transporting tent group equipment, for carrying weapons and ammunition, as a firing platform, and for the evacuation of casualties.

3-4. PACKING

When packing the ahkio, the load must be evenly distributed. Place heavy equipment on the bottom and slightly to the rear of center. Loading lighter equipment toward the top will prevent the ahkio from becoming top heavy. Tools such as shovels, axes, and saws are packed at the sides for easy access in trailbreaking; however, caution must be taken to ensure that these items do not stick out. The canvas cover is folded over the load. To keep the load from shifting and to prevent snow from getting under the cover, lash the load tightly by crisscrossing the lashing rope from the lashing rings on one side of the sled to the rings on the other side. An additional ahkio, issued on a basis of one per two 10-man tents, should be used to carry camouflage nets and support systems. In very wet or dry snow, the movement of an ahkio is easier when wax is applied to the running surface.

3-5. TEN-MAN ARCTIC TENT

The 10-man arctic tent with liner is a six-sided pryamidal tent supported by a telescopic pole. The tent is designed to accommodate 10 men and their clothing and equipment. It will hold additional personnel by leaving packs and equipment outside the tent, or by lowering the center pole, which allows the sidewalls to spread out to cover more ground surface. The tent has two doors, which permits joining of tents for access from one tent to another. A snow cloth is attached to the bottom of the sidewalls for sealing the tent to the ground. The snow cloth should always be to the outside of the tent with pine or spruce boughs underneath, or it may freeze to the ground. This allows easy escape in case of an emergency if the doors are blocked. The tent is ventilated by four built-in ventilators on opposite sides and near the top of the tent.

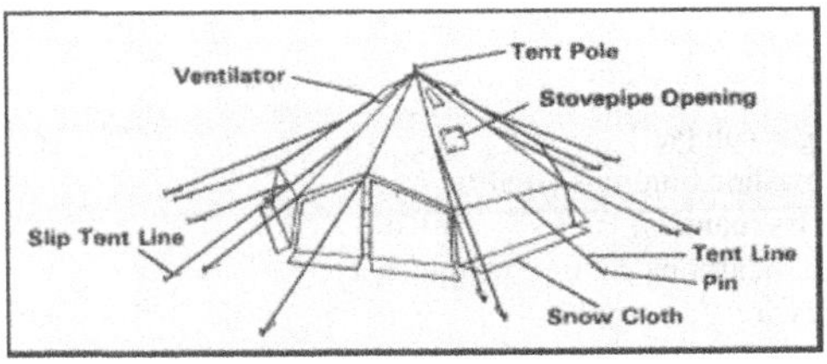

Figure 3-1. Ten-man arctic tent.

3-6. BUILDING ARCTIC TENTS

The shelter is easily built by four trained soldiers. Consider the following when pitching or striking tents in snow or on frozen ground.

Snow should be cleared to the ground surface to obtain a lower silhouette and gain advantage of ground temperatures, which are generally warmer than upper air temperatures. Enough pine or spruce boughs should be placed on the tent and floor, and under the snow covers, for insulation and comfort. If an insulating material is not used, the heat of the stove will thaw the ground insulation in a sea of mud. The snow covers will freeze to the ground, preventing escape from the tent in case of an emergency. When it is impractical to remove the snow to the ground level, an adequate tent site can be prepared by packing the snow with skis or snowshoes until a firm base is provided for pitching. The center pole is placed on a log or other suitable support to keep it from sinking into the snow.

In open terrain, with a strong wind, it will be necessary to build a 1-meter-high snow wall on the windward side of the tent to protect it from the wind. The tent is pitched with the entrance angled about 45 degrees from the downwind side. When the tent is set up, the snow cloth around the bottom of the tent should be spread out so that snow or logs can be placed on it to help in anchoring the tent.

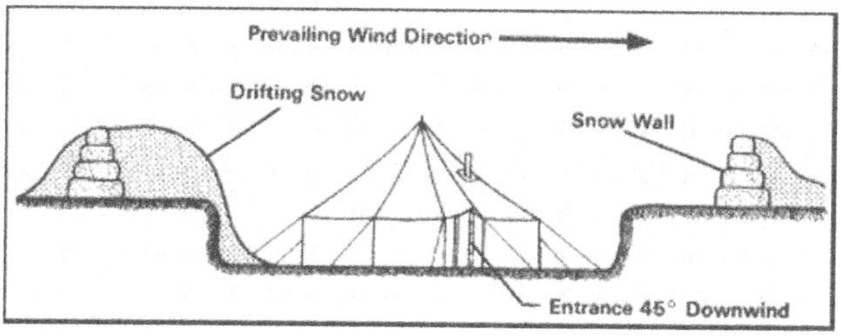

Figure 3-2. Windbreak.

Driving tent pins into frozen ground is impossible. Tents may be pitched and anchored by attaching the tent lines to trees, branches, logs, or stumps. If these are not available, holes are dug into the snow for the purpose of using "deadmen." This is done by digging a hole into the snow large enough to insert a pole or log about 1 meter long with the tent line attached. The hole is filled with snow and well packed. The packed snow freezes, securely anchoring the tent. In powder or dry snow, water would be poured into the hole.

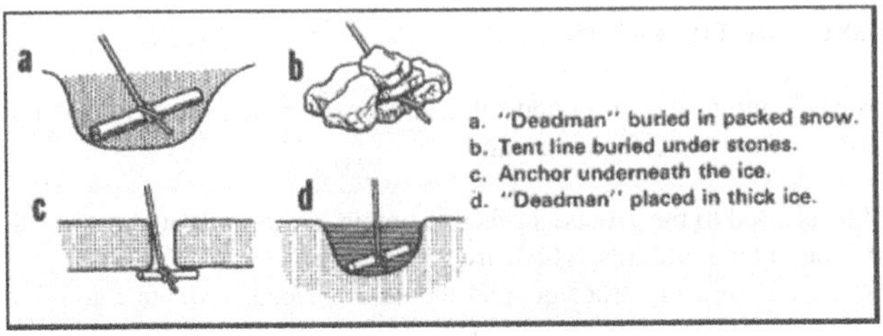

Figure 3-3. Deadmen.

The smaller tent (five-man) provides the commander more flexibility in distribution of his forces. You can never be assigned a tactical mission in the arctic without provisions for a shelter. This tent enables a smaller unit to be assigned an independent mission, such as patrol or reconnaissance mission. It is also six-sided, pyramidal, and supported by a telescopic tent pole. It is designed to accommodate four to five men and their clothing and equipment. The tent can provide shelter for a rifle squad or other similar unit when rucksacks are placed outside the tent.

3-7. INSTRUCTIONS FOR PITCHING TENTS

Instructions for pitching tents are sewn securely to the tent just inside the door. Although the instructions are easily read and followed, there is one thing that should be remembered when pitching either tent. After the tent is spread out flat on the ground, the six corner eave lines should be staked out and loosely secured to the stakes before the tent is raised. These lines are yellow for easy identification. The stakes or tie-down points for these lines should be located about 2 meters from the tent. After this is done, proceed with the tent pitching as indicated in the instructions. The door eave line should be propped up by placing the line over a pole (tree branch or other object higher than the door entrance) and then stake the line out to a pike or tie-down point. This procedure keeps the door from sagging and makes the zipper work better. Drying lines should be attached to the liner inside the tent to allow drying of clothes.

In winter the tent will normally be covered with ice and snow, which must be removed before the tent is struck. This can be done by shaking the tent or by beating it with a mitten or stick. If the snow cloth is frozen to the ground, the snow and ice around it must be removed by chopping or shoveling. This must be done carefully to avoid damage to the snow cloth. One way to do this is to ease the shovel between the cloth and the ground and gently pry the cloth away from the ice.

When the bottom of the tent is properly sealed and the door is zipped, moisture will form on the inside of the tent and collect on clothing and equipment. Carbon monoxide and fumes from the stove can accumulate to a dangerous degree. To avoid this, ensure that the ventilators near the peak of each tent are open.

The roof of the shelter must be steep enough to shed snowfall. With both the 5-man and 10-man tent, the area that will become the floor of the tent is dug, packed, and insulated with pine, spruce boughs, or similar material. Snow walls should then be built to protect against wind.

3-8. YUKON STOVE

The Yukon stove is used to heat the 10-man, 5-man, and GP small tent. In addition to providing heat, the top surface of the stove and the area beneath the stove may be used to heat rations and water. The stoves use standard MOGAS but may also be operated with diesel fuel or JP-4 without modification. NEVER MIX FUELS. When solid fuels (wood, coal, etc.) are used, the stove must be modified by removing the burner unit, closing the burner opening, and turning over the wire grate so there is space below for draft and ashes.

WARNING

DO NOT USE A SPARK ARRESTER ON THE M1950 YUKON STOVE.

A flue cap, issued as a component part of the Yukon stove, shields the stovepipe top from the wind and prevents the wind from blowing smoke or gases down the pipe into the stove and tent. The flue cap is equipped with three 4 1/2-meter lines that are used to steady the stovepipe and anchor it in high winds. These guy lines must be anchored to the tent or tent ropes, not to the ground or nearby trees. The stovepipe hole cover must be rolled up and tied off to prevent nylon ties from touching the stovepipe.

A firm base for the stove must be provided to keep it from melting down into the snow. The base should be 35 centimeters by 70 centimeters and be made of lightweight metal or salvaged plywood. If plywood is used, it should be covered with metal or aluminum foil. If those materials are not available, a base can be made out of small logs about 10 centimeters in diameter.

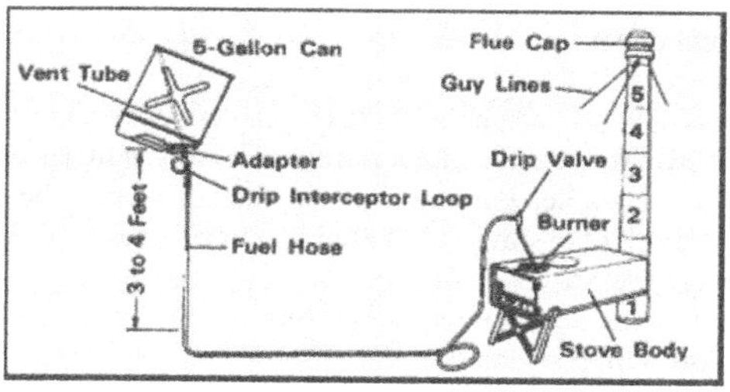

Figure 3-4. Yukon stove.

A simple stand for holding the fuel can is built by obtaining three poles 2 meters long. The poles are tied about two thirds of the way up and then spread out to form a tripod. The fuel can should be at least 1 meter higher than the stove. If the fuel can is wobbly or if there is a wind, the can should be tied to this tripod. Make sure the can is tilted so that the air is trapped in the uppermost corner.

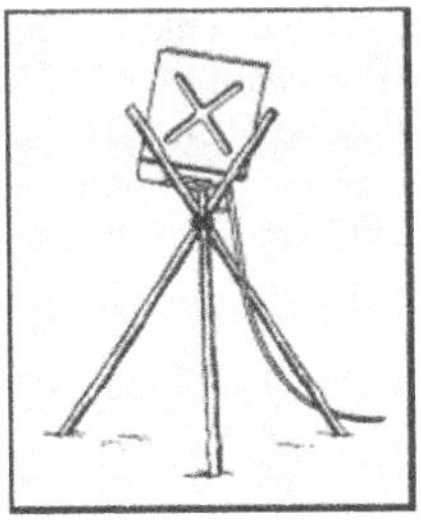

Figure 3-5. Tripod for gas can.

3-9. PRECAUTIONS

The following precautions must be observed when the Yukon stove is being used:

- All stovepipe connections must be tight, and the tent stovepipe shield properly adjusted. The stove must be level to ensure that the burner assembly will spread the flame evenly throughout the stove.
- Ensure that the fuel line is made of natural rubber and suitable for extreme-cold operation. It must be protected so it cannot be accidentally pulled loose and must have a drip intercept loop to prevent fuel from running down the line and into the tent.
- Rate of fuel flow must be checked regularly. It will change with the fuel level and will need adjustment. All fuel supplies must be kept outside the tent.
- The stove should never be left unattended when it is burning. Allowing the stove to become too hot may cause the stove body to become overheated and warped and burn the tent down.
- If the flame is accidentally put out, or if the fuel can is being changed, the drip valve must be turned off. When the stove has cooled, excess fuel inside the stove must be removed. Two or three minutes must pass so gas fumes escape before relighting the burner. When burning gasoline, the burner must be cool before relighting the stove. If the stove is lit before the burner has cooled, the fuel will vaporize, causing it to explode.
- If the tactical situation does not require personnel to be awake 24 hours a day, turn the stove off during sleeping time.

The following are precautions when **burning wood or coal**:

- Fuel should be fed in small amounts until there is a hot bed of coals.
- The stove should not become overheated.
- Gasoline should not be poured on the wood or coal inside the stove to hurry starting the fire.
- Ashes should be removed regularly and not allowed to accumulate higher than the bottom of the grate.
- NEVER MIX FUELS.

3-10. SQUAD STOVE, M1950

The one-burner squad stove is a cooking and heating unit for two to five men. It is used when operating in an isolated or forward area when separated from the ahkio group. The stove is small, compact, and lightweight, and it will operate on either MOGAS or white gasoline. The initial pressure for its operation is generated by a few strokes on the hand pump. Thereafter, the pressure is generated by heat from the burner evaporating the fuel in the stove tank.

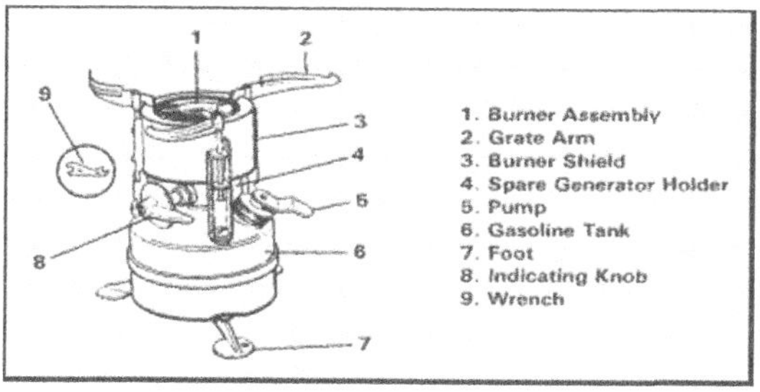

Figure 3-6. Gasoline cooking stove, M1950.

To operate the squad stove, do the following:

- Snap out the three feet and set stove on a level surface.
- Raise the grate arms and turn them outward so that the inner tabs can be engaged in the slots in the burner shield.
- Close the valve by turning the indicating knob clockwise to the OFF position.
- Unscrew the pump caps and remove the pump assembly.
- Fill the gasoline tank three-fourths full.
- Wipe off any gasoline that may have been spilled during filling and replace the pump cap.
- Pump 10 strokes to build up initial pressure.
- Turn the indicating knob to the ON position for 3 seconds. This allows the gasoline to rise through the generator to the burner and preheated cup.

- Turn the indicating knob to the OFF position to prevent excess fuel from accumulating in the burner assembly.
- Light the wick and allow the flame to burn for about 3 minutes or until the flame turns blue.
- Turn the indicating knob to the LIGHT position. Have a second match ready to relight the burner immediately if the flame goes out.
- When the flame in the burner turns blue, turn the indicating knob to the ON position.
- Pump an additional 10 strokes to maintain pressure within the tank. The stove is now ready for use.
- If the flame burns unevenly or dies during operation, turn the valve to the OFF and ON position several times, relight quickly, and pump several additional strokes.
- To turn the stove off, turn the indicating knob to the OFF position and allow the flame to go out.

CHAPTER 4

Small Arms and Ammunition

"Rifles and machine guns and even the breechblocks of artillery became absolutely rigid ... and steel parts cracked."

General Rendalic, Eastern Front, 1942

4-1. GENERAL

Cold weather will greatly affect the operation of your weapon. You must be aware of these effects to ensure employment of your weapon under poor weather conditions.

4-2. CARE, CLEANING, AND MAINTENANCE

Your weapon will work properly under extreme cold conditions if you take care of it. Normal temperate zone lubricants thicken in cold weather, and stoppages or sluggish actions of your weapon will result. During the winter, disassemble your weapon completely, clean it thoroughly, and remove all lubricants. Oil the entire weapon with a light coating of cleaner lubricant preservative (CLP).

CAUTION
DO NOT USE CLP ON A MORTAR OR ANY COMPONENT PART OF A MORTAR.

Figure 4-1. Weapons will 'sweat" when taken in to a heated shelter.

A main problem is to keep snow and ice, which may disable your weapon, out of the working parts, sights, and barrel. When the weapon is not in use, keep it out of the snow either in an improvised weapon stand or in a tripod arrangement. If this is not possible, stand it up with the butt end in the snow.

Condensation ("sweating") will form on the weapon when you take it from the cold into a heated shelter. The condensation will continue for about 1 hour, after which you should clean the weapon. If you don't, the condensation will freeze when you take the weapon outside, and it may not fire. The weapon must be kept as near the floor as possible to lessen condensation. If parts of the weapon do become frozen, warm them and move the working parts slowly until they are unfrozen. If you can't warm your weapon, remove all ice and snow and move the parts gently until action is restored. The use of CLP will prevent malfunctions by eliminating condensation.

When firing, do not let the hot parts of your weapon come in contact with the snow. The snow will melt and, as the weapon cools, will form ice. When changing barrels, don't lay them on the snow.

To prevent bipods and tripods from sinking into the snow, improvise for added flotation. This is done by placing them on snowshoes, ski poles, branches, ahkios, etc. Baseplates should be on solid ground or sandbags to prevent sinking into the snow and damage to the weapon.

Figure 4-2. Bipods and tripods will sink in the snow.

AMMUNITION

Small arms ammunition should be kept at the same temperature as your weapon. Ammunition clips and magazines must be cleaned of all oil and preservative and checked often. All ice, snow, and condensation must be removed. Ammunition boxes must be kept closed in order to keep out snow and ice.

Stored ammunition should be kept in packed boxes, raised off the surface by placing them on boughs, cardboard, or other similar material, and covered to protect them from the snow. Ammunition should be well marked for easy location and identification in case it becomes covered with snow.

Ammunition resupply may become restricted. Everyone must be aware of the necessity for ammunition economy and fire discipline. Loaded clips, magazines, or single rounds dropped in the snow become quickly lost; therefore, careful handling of ammunition is essential. Never waste your ammunition. Make every round count.

When contact with the enemy is imminent, troop compartments and aircraft should be kept as close to freezing as possible to avoid condensation. However, many munitions have limitations at lower temperatures, which must be recognized in training.

CHAPTER 5

Rations and Diet

"I ate my boots and a pair of pants ... tea and seal-skin gloves for dinner."

Greely, Arctic Expedition, 1883

5-1. GENERAL

Most of what we eat and drink is used in maintaining our body heat, while only a small portion is used in producing energy for physical work. You must ensure adequate caloric intake in cold-weather operations. About 4,000 calories per day are necessary for personnel performing physically demanding work in the cold. Efficiency may drop rapidly if this level is not maintained.

The body loses liquid at a very fast rate in arctic conditions, regardless of how carefully you adjust and ventilate your clothing. The exertion of movement on foot, preparation of bivouac sites and defenses in the snow, etc., take a toll in sweat and loss of moisture in the breath. These liquids must be regularly replaced, preferably by hot drinks, which provide extra calories if they contain sugar.

5-2. RATIONS

Rations provide the needed calories to live and fight effectively. When eaten in their entirety, rations contain the right amount of carbohydrates, fats, protein, and vitamins. The proper intake of these essential items depends on the entire ration being eaten in properly spaced meals. The lack of concern regarding eating caused by the cold, combined with the difficulties and inconvenience of cooking, may tempt soldiers to miss meals. The principles of sound leadership and discipline in cold weather require that meals be prepared and that the entire ration be eaten, warmed when possible. Save snacks (cookies, chocolate bars, etc.) for between meals and when on the march.

5-3. LIQUIDS

Because there is a large amount of fluid loss in the cold, you should drink a minimum of 3.5 quarts of water a day when heavy physical activity is involved. Water is usually available either from streams or lakes or by melting snow or ice. A limiting factor may be fuel needed to melt the snow or ice. To save time and fuel, water should be obtained from running streams or a lake. The milky water of glacial streams must be allowed to stand until the sediment settles. When a hole is cut into the ice to get water, it should be covered by a snow block or something similar to slow refreezing.

Ensure that as much of the daily liquid intake as possible is hot drinks, such as soups and cocoa. Your local command may request a special food allowance of warming and cooling beverages. Main meals should begin with soup, and between-meal snacks should include a hot drink.

When running water is not available, ice or snow must be melted. Ice produces more water in less time than snow. When melting snow, first put a small amount into the cooking pot; then more can be added. Continue this process until there is enough water. It must be purified by boiling rapidly for 15 minutes.

When heating water, use all available stoves, since this is a most time-consuming operation. Melting and boiling enough snow for a drink may take 30 to 40 minutes. Areas designated for ice or snow that is to be used for water must be sited far upwind of the latrine and garbage disposal sites. Before going to bed, ensure that stoves are filled and enough water has been prepared for breakfast. Do not let other soldiers eat snow or ice; doing so lowers the body's core temperature.

Never drink alcohol on operations or when exposed to low temperatures. The aftereffects of alcohol can lead to a feeling of false security. You may forget the main rules to prevent cold injury, like wearing your gloves or not going outside without shoes to relieve yourself.

CHAPTER 6

Hygiene and First Aid

"...the first frostbite cases have made their appearance. We amputated two feet and will amputate many more."

Amosoff

6-1. GENERAL

The care of your body needs special attention in cold weather. Ensuring cleanliness in the field, particularly in the cold, is important. It is easy to let yourself go for long periods without bathing, but if you wish to be part of an effective combat force, you must keep clean.

6-2. HYGIENE TIPS FOR THE SOLDIER

Face, hands, armpits, and crotch should be washed daily. If water is not available, take a "bath" using cornstarch or rub your body briskly with a dry towel. The cornstarch removes excess oil and perspiration from skin and hair.

Shave regularly -- daily if water is available or every 2 to 3 days if water is in short supply. Shaving just before sleeping will allow the face maximum recovery time before going out into the cold, reducing the possibility of superficial frostbite.

Teeth should be cleaned daily. If a toothbrush is not available, a clean piece of cloth wrapped around a finger or a chewed twig may be used as a substitute.

Underwear should be changed at least twice a week. If it is not possible to wash your underwear, crumple it, shake it, and air it for at least 2 hours.

Ensure that all eating and cooking utensils are cleaned after each use.

In the dry arctic you will find the need to oil your unprotected and exposed skin once every week (with an oil such as suntan oil).

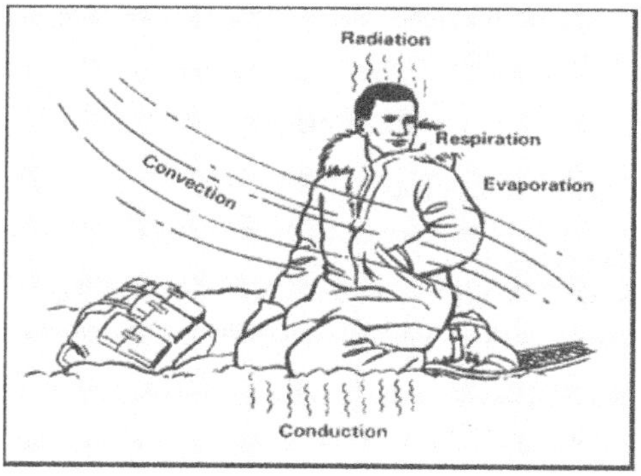

Figure 6-1. Mechanisms of heat loss from the body.

6-3. HYPOTHERMIA

Hypothermia is a lowering of the temperature of the body's inner core. This happens when the body loses heat faster than it can produce it. You must be able to recognize the symptoms of hypothermia and to treat the condition immediately.

PREVENTION

To prevent hypothermia, adhere to the following:

- Stay physically fit.
- Keep active.
- Use the uniform properly and keep it dry.
- Eat properly and often.
- Drink plenty of liquids, at least 3.5 quarts per day, when performing physically demanding work.
- Be prepared for and know how to deal with quick changes in the weather.
- Bivouac early before judgment is weakened by fatigue.

SYMPTOMS

Be aware of the following symptoms:

- Shallow breathing or absence of breathing.
- Faint or undetectable pulse.
- Victim is first cold, then stops shivering.
- Loss of feeling.
- Poor coordination; victim has trouble walking.
- Impaired speech.

- Confusion, withdrawn appearance, depressed, and/or uncoordinated. An uncaring attitude and/or glassy stare are also clues.

TREATMENT

The recommended treatment is as follows:

- Prevent any further heat loss.
- Get the victim out of the wind and into the best shelter available.
- Replace any wet clothing with dry, and place the victim in a sleeping bag, if one is available.
- Place as much insulation as possible between the victim and the ground.
- Add heat by the best available means to the victim's neck, groin, and sides of chest. This can be done by using a hot water bottle, Yukon stove, campfire, or your own body heat.
- Provide the victim with something warm and nutritious to drink if he is conscious. Calories may be added by using sugar in hot, sweet drinks.
- DO NOT massage the victim.
- DO NOT give alcohol to the victim.
- Evacuate the victim to the nearest medical treatment facility as soon as possible. This condition is life threatening until normal body temperature is restored.

6-4. FROSTBITE

Frostbite results when tissues freeze from exposure to temperatures below 32°F. The degree of injury depends upon the wind-chill factor, length of exposure time, and proper protection. Persons with a history of cold injury are prone to frostbite. There may be a tingling, stinging, an aching sensation, or a cramping pain. The skin first turns red and then becomes pale gray or waxy white. Frostbite can be classified as superficial or deep.

WIND CHILL

The combination of wind and low temperatures creates a condition known as wind chill. For example, with the wind calm and a temperature of -20°F there is little danger of wind chill. If the temperature is -20°F and there is a wind of 20 knots, the equivalent chill temperature is -75°F!

Under these conditions there is GREAT DANGER and your exposed skin can freeze within 30 seconds. You also create your own wind when you walk, run, or ski. As examples, walking creates a wind speed of about 5 mph and running 10 mph. Wind is also created by moving vehicles, aircraft propellers, and rotors. The wind-chill chart is provided to show how wind and temperature combine to create wind chill; use it. Cold injury can disable you like an enemy bullet.

WIND SPEED		COOLING POWER OF WIND EXPRESSED AS "EQUIVALENT CHILL TEMPERATURE"																				
KNOTS	MPH	TEMPERATURE (°F)																				
CALM	CALM	40	35	30	25	20	15	10	5	0	-5	-10	-15	-20	-25	-30	-35	-40	-45	-50	-55	-60
		EQUIVALENT CHILL TEMPERATURE																				
3-6	5	35	30	25	20	15	10	5	0	-5	-10	-15	-20	-25	-30	-35	-40	-45	-50	-55	-60	-70
7-10	10	30	20	15	10	5	0	-10	-15	-20	-25	-35	-40	-45	-50	-60	-65	-70	-75	-80	-90	-95
11-15	15	25	15	10	0	-5	-10	-20	-25	-30	-40	-45	-50	-60	-65	-70	-80	-85	-90	-100	-105	-110
16-19	20	20	10	5	0	-10	-15	-25	-30	-35	-45	-50	-60	-65	-75	-80	-85	-95	-100	-110	-115	-120
20-23	25	15	10	0	-5	-15	-20	-30	-35	-45	-50	-60	-65	-75	-80	-90	-95	-105	-110	-120	-125	-135
24-28	30	10	5	0	-10	-20	-25	-30	-40	-50	-55	-65	-70	-80	-85	-95	-100	-110	-115	-125	-130	-140
29-32	35	10	5	-5	-10	-20	-30	-35	-40	-50	-60	-65	-75	-80	-90	-100	-105	-115	-120	-130	-135	-145
33-36	40	10	0	-5	-15	-20	-30	-35	-45	-55	-60	-70	-75	-85	-95	-100	-110	-115	-125	-130	-140	-150
WINDS ABOVE 40 HAVE LITTLE ADDITIONAL EFFECT		LITTLE DANGER					INCREASING DANGER (Flesh may freeze within 1 minute)						GREAT DANGER (Flesh may freeze within 30 secs)									

Figure 6-2. Wind-chill chart.

PREVENTION

Frostbite is preventable if the individual soldier takes care of himself. It seldom occurs in people who are maintaining enough body heat. It is most commonly associated with an overall body heat loss resulting from poor equipment, reduced food intake, dehydration (water loss), overtiring, injury, or a combination of these factors. The feet, hands, ears, and exposed areas of the face are most easily hurt by frostbite and must receive constant care. The following are rules to help prevent frostbite:

- Do not wear tight boots and socks.
- Use the "buddy system" to check exposed areas, especially when wind is present.
- Carry extra socks and mitten liners.
- Do not stay still for long periods.
- Use caution when cold and wind are combined.
- Check feet during halts.
- Ensure that you eat well and have hot drinks as often as possible.

SYMPTOMS OF SUPERFICIAL FROSTBITE

Redness occurs, followed by powdery flaking of the skin. Affected areas of dark-skinned soldiers may appear dull and grayish. Blister formation occurs 24 to 30 hours after exposure, followed by flaking of superficial skin.

SYMPTOMS OF DEEP FROSTBITE

Common symptoms are as follows:

- Loss of feeling occurs in the affected area, leaving it pale yellow and waxy looking.
- When the frostbitten area thaws, it is painful.
- Frozen tissue may feel solid or "wooden" to the touch.
- When exposed to inside temperatures, blisters may appear in 12 to 36 hours.

- Discoloration (red-violet) appears 1 to 5 days after the injury.
- Gangrene usually results.

TREATMENT FOR FROSTBITE

Determine whether the frostbite is superficial or deep. If the exposure time was short, the frostbite will probably be superficial. If the exposure time was long, the frostbite will probably be deep.

The following is a recommended treatment for frostbite:

- Move the casualty to a warm and sheltered area.
- Do not rub the frostbitten area with snow or apply cold water soaks.
- Do not rewarm the affected area by massaging or exposing it to open fire.
- Rewarm the face, nose, or ears by placing your hands on the frozen area. Rewarm frostbitten hands by placing them under clothing and against the body.
- Close the clothing to prevent further loss of body heat.
- Rewarm the feet by removing the boots and socks. Place the bare feet under the clothing and against the abdomen of a buddy. Once the feet are warmed, put on dry socks and boots, if available. If the victim must wear the wet socks and boots, he should exercise his feet by wiggling his toes.
- Loosen tight clothing and remove jewelry.
- Improve circulation by exercise.
- Do not allow the victim to use alcohol or tobacco. Alcohol increases loss of body heat; tobacco causes narrowing of blood vessels in the arms and legs.
- Reassure the victim.
- Protect frozen tissue from further cold or trauma, and evacuate the casualty to the nearest medical treatment facility.

Deep frostbite is most common and harmful to the feet and less common to the hands and ears. When deep frostbite does occur, the victim should be moved to a sheltered area and evacuated to a medical treatment facility immediately. If possible, do not allow the victim to walk if his feet are frozen and avoid treating or thawing the affected area. Thawing of deep frostbite in the field increase pain and invites infection, greater damage, and gangrene. Do not try to thaw deep frostbite; let the physician do it.

6-5. DEHYDRATION

Dehydration can cause serious physical problems and make you more susceptible to other problems, such as frostbite and hypothermia.

PREVENTION

The following are suggestions to prevent dehydration:

- The **minimum** daily liquid requirement for personnel performing demanding physical work in the cold is 3.5 quarts per day.
- By the time you feel thirsty you are already dehydrated. Drink whenever possible, particularly during halts.
- Ensure that canteens are full before any type of movement.
- During periods when water is scarce, do not drink coffee in large quantities. It will pass through you quickly. Water should be the first choice.
- Check your urine spots in the snow. Dark yellow or brown indicates dehydration.

SYMPTOMS

Common symptoms are as follows:

- Red indicates a severe condition needing immediate medical evacuation to the nearest medical treatment facility.
- Lack of appetite.
- Dry mouth, tongue, and throat.
- Stomach cramps and/or vomiting.
- Headaches.

TREATMENT

Recommended treatment is as follows:

- Keep the victim warm.
- Give him plenty of liquids.
- Make him rest.

6-6. CARBON MONOXIDE POISONING

Whenever a stove, fire, gasoline heater, or internal combustion engine is used indoors, there is danger of carbon monoxide poisoning. Fresh air in living and working quarters is vital. Carbon monoxide is a deadly, odorless gas.

PREVENTION

The following are suggestions to prevent carbon monoxide poisoning:

- Use stoves and lanterns in well-ventilated areas and tents.
- Ensure that stoves and lanterns are functioning properly.
- Do not let personnel warm themselves by engine exhaust.
- Always have windows cracked in vehicles with a heater in use.

- Use a tent guard or shut the stove off when sleeping.

SYMPTOMS

Common symptoms are as follows:

- Headache, dizziness, confusion, yawning, weariness, nausea, and ringing in the ears.
- Bright red color on lips and skin.
- Victim may become drowsy and collapse suddenly.
- If personnel are found unconscious in an enclosed shelter, carbon monoxide poisoning should be suspected.

TREATMENT

Adhere to the following for proper treatment:

- Move the victim to open air.
- Keep the victim still and warm.
- If the victim is not breathing, administer mouth-to-mouth resuscitation.
- Administer cardiopulmonary resuscitation, if necessary.
- Immediately evacuate the victim to the nearest medical treatment facility.

6-7. SNOW BLINDNESS

Snow blindness is caused by ultraviolet and ultrablue rays of the sun being reflected from a snow-covered surface into the eyes. This condition can occur even in cloudy weather. In fact, it is more likely to occur in hazy, cloudy weather than in sunny weather.

Figure 6-3. Snow blindness.

PREVENTION

Prevention is easy; wear your issued sunglasses. If sunglasses are not available, make slitted glasses from cardboard, thin wood, tree bark, or similar material.

SYMPTOMS

Common symptoms are as follows:

- A scratchy, sandy feeling under your eyelids.
- Redness and watering of the eyes.
- A headache.

TREATMENT

Adhere to the following for proper treatment:

- Blindfold the patient using a dark cloth.
- Reassure the patient.
- Evacuate the patient to a medical treatment facility.

6-8. SUNBURN

Sunburn is a serious hazard because of the reflective qualities of the snow, especially at higher elevations.

PREVENTION

Require soldiers to use a sunscreen and lip balm.

SYMPTOMS

Redness of skin with slight swelling is one symptom of sunburn. Prolonged exposure to the sun may cause pain and blistering and, in severe cases, chills, fever, and headaches.

TREATMENT

Soothing skin creams may be helpful if the swelling is not severe.

6-9. TENT EYE

Tent eye is an inflammation of the eye caused by fumes from stoves and lanterns used in a poorly ventilated tent or shelter. It is prevented by proper ventilation of tents or shelters. The treatment is fresh air.

6-10. TRENCH FOOT/IMMERSION FOOT

Trench foot and immersion foot are injuries caused by the prolonged exposure of skin to cold or wet conditions. In extreme cases, the skin dies and amputation of the foot or leg may be necessary.

PREVENTION

The following helps prevent trench foot or immersion foot:

- Change to dry socks at least twice daily.
- Dry and massage your feet regularly with foot powder.
- Wear VB boots and change socks regularly.
- If leather boots are worn, dry them whenever possible.

SYMPTOMS

In early stages, the feet and toes are cold, numb, and still, and walking becomes challenging. The feet swell and become painful.

TREATMENT

In early stages, keep the feet dry, clean, and exposed to the air. In later stages, evacuate the victim to a medical facility.

6-11. CONSTIPATION

At very cold temperatures, personnel tend to put off the natural urge to have a bowel movement, resulting in constipation. This can be a painful casualty producer.

PREVENTION

Eat fresh or canned fruit and eat regularly. Drink plenty of liquids. Try to have a daily bowel movement at your normal time.

SYMPTOMS

Some common symptoms are stomach cramps, dizziness, and headaches.

TREATMENT

Prevention is the best treatment. If symptoms persist, get medical help.

6-12. DIARRHEA

Diarrhea can be treated by drinking tea leaves with water.

6-13. HEAT CRAMPS

Heat cramps are caused by an excessive loss of salt in the body. This loss of salt causes muscles to spasm, and they are unable to relax. After prolonged physical activity in a cold-wet environment, the signs and symptoms of heat cramps could appear in the arms, legs, and/or stomach. It is possible to experience heat cramps in cold-weather environment when undergoing extreme physical exertion.

SYMPTOMS

The soldier experiences muscle cramps in his arms, legs, and/or stomach. He is pale and has wet skin and is experiencing dizziness and extreme thirst.

TREATMENT

Have the soldier drink one fourth of a canteen of water. Give the soldier one canteen of water with one packet of salt stirred in. Have him drink the canteen of salted water over a 30-minute period. If cramping of muscles continues, evacuate the soldier to the nearest medical treatment facility.

NOTE

Do not give the saltwater to a soldier who has an upset stomach. Have him drink a canten of unsalted water.

6-14. HEAT EXHAUSTION

Heat exhaustion is caused by dehydration and loss of body salt from extreme physical activity in a cold-wet environment. It is possible to experience heat exhaustion in a cold-weather environment when undergoing extreme physical exertion.

SYMPTOMS

The soldier may feel dizzy, weak, and/or faint. The skin feels cool and moist to the touch. He may feel nauseated or have a headache.

TREATMENT

The following is a recommended treatment for heat exhaustion:

- Have the soldier drink one canteen of water to relieve the symptoms.
- If he complains of cramps, give him one canteen of water with one packet of salt stirred in. Have him drink the salt solution over a 30-minute period.

NOTE

Do not administer salt if the soldier has an upset stomach. Have him drink a canteen of unsalted water.

- Loosen all tight-fitting clothes.
- Elevate the feet above the level of the heart.
- If the soldier is unable to drink water due to an upset stomach or if the symptoms have not improved within 20 minutes, evacuate the soldier to the nearest medical treatment facility.

CHAPTER 7

Bivouac Routine

> ***"Had we lived, I should have a tale to tell of the hardihood, the endurance and the courage of my companions which would have stirred the hearts of every Englishman. These rough notes and our dead bodies must tell the story."***
>
> ***CPT Robert Scott***

7-1. GENERAL

The essential requirements for survival and mission accomplishment in the cold/arctic environment are warmth, food, and shelter. As covered in Chapter 3, you are provided with tents that can be erected quickly. Tents can be lost or destroyed and separated from the ahkio groups in the confusion of combat. When this occurs, timber and snow, which are usually available, will be used for shelters. A detailed explanation of the bivouac routine should be contained in the unit SOP.

7-2. BIVOUAC SITE

Bivouac sites, other than those that are permanent nature in rear areas, either are tactical or are designated to provide living accommodations within a defended position. The tactical bivouac is usually not occupied for more than one night, is established at or after last light, and needs little preparation. It is organized for a rapid and disciplined response to enemy action. The defensive position is likely to be occupied for longer periods and must be established accordingly. The layout and placement of tents and shelters must be coordinated within the defense and in accordance with the unit SOP.

7-3. SITE SELECTION

The selection of bivouac areas is important and needs careful consideration. The challenge of area selection varies with the tactical situation, weather conditions, and terrain. The bivouac area will be tactically located in accordance with the principles of security and defense.

There must be concealment from air and ground observation and wind. Forested areas pose few challenges in comparison to areas that are located above the tree line. In selecting areas, ensure that local camouflage materials are available for use with the snow camouflage nets. Areas that offer better bivouac sites than others are explained below.

FORESTED AREAS

Forests provide excellent sites. Camouflage is easy and building materials, firewood, and wood for insulation are readily available. Although forests provide the best bivouac site, temperatures there will normally be colder than those in open areas.

MARSHY GROUND

In winter, good sites are frozen swampy areas, the banks of streams, and the shores of lakes. Some swampy areas may not freeze because of warm springs or marsh gases. Deep valleys should be avoided since colder air gathers there.

OPEN AREAS

Avoid open areas because of strong winds, drifting snow, and poor concealment. If avoidance is not possible, then pitch tents where they can be sheltered by natural windbreaks. These are normally found in depressions or behind ridges of ice on lakes. In areas where natural wind falls do not exist, build snow walls to provide wind protection. In areas with high winds, snow gathers on the lee side, making it necessary to clear the side and tops of the tents, often to prevent the weight of the drifting snow from collapsing the tent.

MOUNTAINOUS AREAS

Mountainous terrain is characterized by strong winds, cold, and a general lack of concealment above the timberline. The wind over head creates extensive lees near cliffs and steep rock faces. Avalanches can occur on steep slopes; avoid their likely path. Signs to look for are:

- Heavy snow overlaying a crust of ice (usually the result of a freeze at the end of a period of thaw).
- New snow on steep slopes if the temperature is rising.
- Evidence of avalanches in previous years; an area cut through the trees (branches stripped off, and/or broken tree stumps at different heights) or a smooth clear strip of ground down the mountainside.

Figure 7-1. Location of tent.

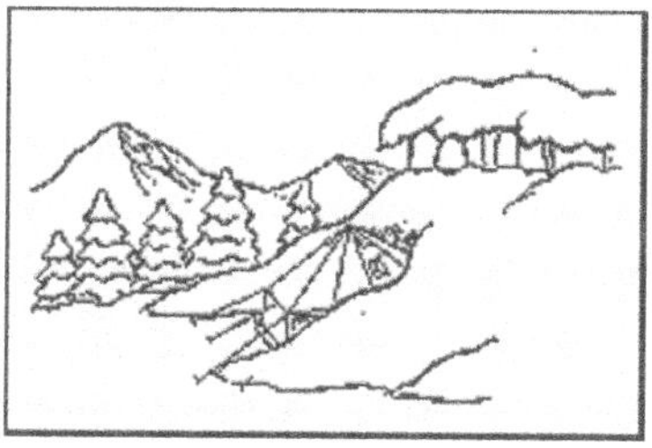

Figure 7-1. Location of tent (continued).

When contact with the enemy is likely, the unit SOP will identify specific considerations when selecting the site for a tactical bivouac. For example:

- The location of the enemy.
- Enemy patrol habits.
- The air threat.
- Dominating features that identify enemy avenues of approach.
- The wind direction that can carry sounds of movement (the bivouac site should be downwind of likely enemy avenues of approach, especially when wind speeds are low).
- Availability of natural camouflage.
- Amount of moonlight and weather conditions. In good weather or bright moonlight, the bivouac site must be in an area of deep shadow (in the middle of the woods or on the northern slope of a ridge) or the tents will be silhouetted.

7-4. RECONNAISSANCE OF BIVOUAC SITE

The main body should be preceded by a reconnaissance party whose task is to lay out the bivouac site. They should:

- Identify the bivouac site and a false position.
- Establish a trail plan.
- Select defensive positions.
- Mark the location of tent sites.
- Identify areas for cutting boughs.
- Act as guides for the main body on its arrival.
- Provide security during occupation by the main body.

7-5. TRAIL PLAN

Trail plans must be established before entering the bivouac site. Normally, the incoming trail will be extended well beyond the area of the bivouac and may lead to a false position. Leading off the incoming trail establish another trail, usually at an extreme angle to entry. The trail network that leads to each subordinate unit area, and within, must follow the inside of the tree line. This prevents enemy observation of trails from the air or ground.

7-6. OCCUPATION

Guides from the reconnaissance party will arrange to meet the main body and identify the layout of the bivouac site. Once the unit is in position, the unit should:

- Identify a temporary area for weapons and equipment during establishment of the site.
- Identify the exact location and type of defensive positions to be built.
- Determine exact tent sites regarding defensive positions, natural shelter, and camouflage.
- Identify sleeping areas and type of shelters to be built when tents are not used.

7-7. BOUGH-CUTTING AREAS

Bough-cutting areas for bedding or improvised shelters should be for all personnel in the unit. These areas must be where activity is not likely — well behind the unit's defensive area or in a decoy bivouac site.

Cut boughs from the bottom portion of trees above the snow and leave the trees standing. Be sure to cut only a few boughs from each tree. A tree stripped of its limbs is easily observed from the ground and from the air.

Cut firewood near the bough-cutting area to allow the same trail to be used. Dead spruce, pine, or squaw wood makes the best firewood. Standing deadwood burns the best and hardwoods the longest. In an arctic and subarctic environment, true hardwoods are rare. Of the trees available, birch, tamarack, and spruce are the hardest woods. Burn the upper parts of dead trees during daylight because they give off lighter colored smoke. The lower parts of the trees contain resin and tar, which burn well but make more and darker smoke. In the daylight this will allow your position to be easily identified by air and ground observation.

7-8. GARBAGE

All garbage should be double bagged and transported to the rear or disposed of in accordance with the unit SOP. Units should never leave evidence of their presence. Carry all trash, if necessary, until it can be disposed of.

7-9. LATRINE

Normally there will be a central latrine within the bivouac site, if the site is not too widely spread. One latrine can usually serve the needs of a platoon-size element. It must be placed downwind of the bivouac site, but not so far that soldiers won't use it. The area identified can be as elaborate as a pit or cross-cut-type latrine that is made windproof by branches, snowblocks, and ponchos. It should be properly camouflaged. It can also be an area where plastic bags are located. Double-bag, tie, and transport all waste to the rear for disposal as identified in the unit SOP. Site urinal areas around a "dedicated tree" or the gasoline tripod, which will hide the discolored snow and will be easily spread about when leaving the bivouac site.

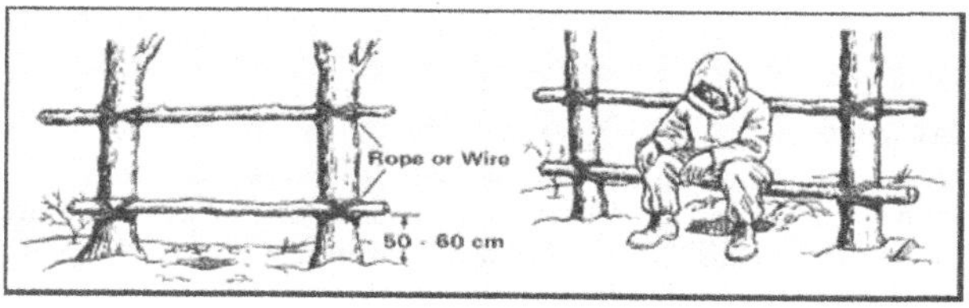

Figure 7-2. Cross-tree type latrine.

7-10. WATER POINTS AND SNOW AREAS

When obtaining water for drinking purposes by melting snow, set aside and restrict an area for this purpose only. The site must be upwind from the bivouac area and isolated from the latrine and garbage areas. The snow must be clean, white, and free of fungus from trees. NEVER use snow that is discolored. Remember — disinfect the water using purification tablets and bring it to a rolling boil for a minimum of 15 minutes.

7-11. STORAGE

Storage problems in the winter are increased by snow, low temperatures, thaws, and limited storage space. Space in any shelter is limited. Store only those items inside the tent that are affected by the cold, or must be immediately available. The remaining equipment should be centralized, well marked, and covered. Store ammunition and fuel separately. Place wood or metal containers on boughs or poles to keep the containers from freezing to the ground. The unit SOP will specify what equipment is placed inside or outside of the tent and where.

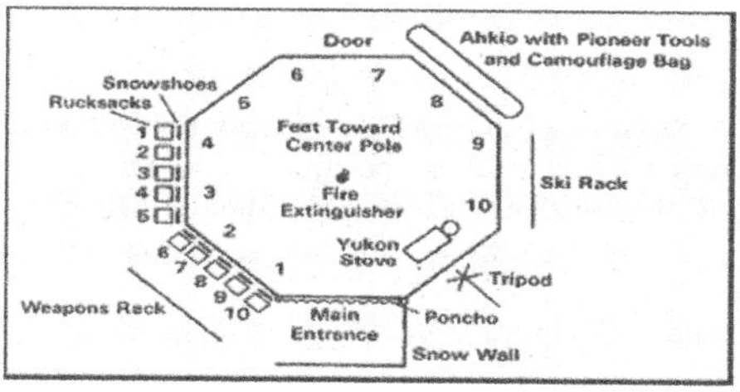

Figure 7-3. Ten-man tent.

CHAPTER 8

Individual Movement

"Anticipate every situation which can arise in battle and think out your reaction to it well in ADVANCE."

Hoote

8-1. GENERAL

Snow affects movement many ways. Heavy snow cover hinders or slows movement, either cross-country or on roads. Snow cover also blankets many terrain features and hides obstacles.

The effect of snow on movement varies greatly with its depth and physical characteristics. Men on foot cannot travel well over flat terrain or roads when the snow is deeper than 30 centimeters. However, movement can be improved by the use of skis or snowshoes.

SKIS	SNOWSHOES
• Faster • Less Work • Harder to Learn	• Slower • More Work • Easy to Learn

Figure 8-1. Using skis or snowshoes.

8-2. SKIS

Skis provide greater speed in moving, particularly over cleared trails. Troops mounted on skis and towed behind small unit support vehicles, or similar vehicles (ski-joring), are effective means for rapid cross-country movement. In heavy vegetation, skiing will always be more physically taxing than snowshoeing.

8-3. SNOWSHOES

Learning how to use snowshoes requires less training. Snowshoe movement is more practical in confined areas, such as assembly positions, field trains, and mortar and artillery positions. If traveling on foot, you should become skilled on both skis and snowshoes.

Your ability to use, and the availability of, over-snow equipment is a major factor in learning to travel well over snow-covered terrain. With previous training in cross-country skiing and if equipment is available, travel on skis is recommended. In most snow conditions and over most types of terrain, skis provide the fastest and most energy-saving form of travel. Use of snowshoes requires little training, but your speed will be slower and travel more exhausting.

Snowshoes are **more** exhausting than skis. However, a light crust on the surface of the snow prevents skis from sinking and provides for fast and easy skiing. A crust hard enough to support a man makes travel on foot feasible, but even then travel on skis is recommended.

In a survival situation, improvise equipment for travel if snow is loose and deep. Make snowshoes of willow or other springy wood using a wood separator and thong, wire, cord, etc.

USE SKIS WHEN:

- Speed is essential.
- Long distance must be traveled.
- Stealth is necessary.
- Conditions allow.

USE SNOW SHOES WHEN:

- Moving through heavy brush.
- Speed is not essential.
- Troops are not proficient on skis.

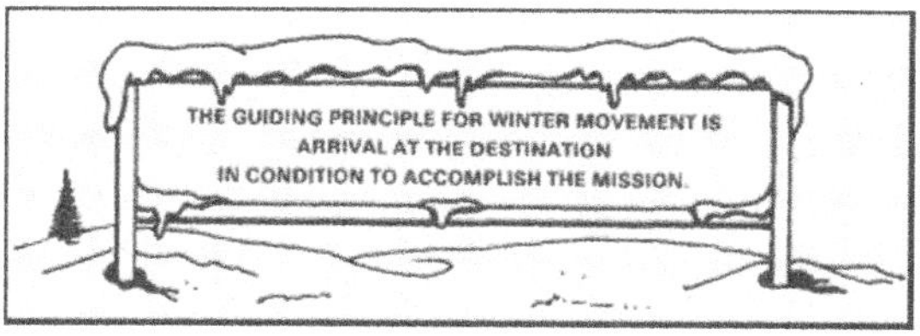

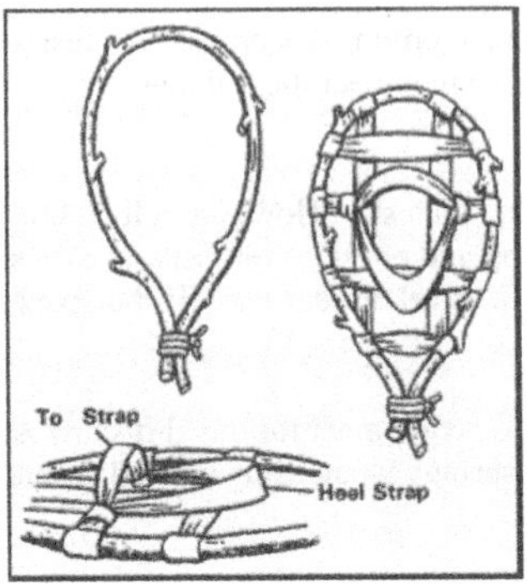

Figure 8-2. Improvised snowshoes.

CHAPTER 9

Land Navigation

"No - I've never been lost; but for about three weeks I was kinda puzzled."

Brigadier General Fergusson

9-1. GENERAL

In some cold-weather environments, accurate up-to-date maps may not be available and landmarks may be few or nonexistent. When the terrain is snow covered, one piece of ground looks like the other. Whatever the circumstances, you must be able to find your way — if not by map reading, by some other means of navigation.

9-2. CHALLENGE OF LAND NAVIGATION

Only poor road nets may exist in some areas. These will be of little use in cross-country navigation. If large-scale maps exist, they may be unreliable.

Long nights, falling or blowing snow, and fog limit visibility in winter. Tracks and landmarks may be erased by heavy snowfall. Accurately identify landmarks, through both ground and air observation, to prevent mistakes.

Conditions of daylight are different from those found in temperate regions (moderate climate). In winter there are long nights and short days.

Magnetic disturbances may be common, thereby making compass readings unreliable. Magnetic declinations may be extreme and can vary considerably at close points.

Many small lakes, ponds, and creeks will be found that are not on the map. The size and location of waterways may vary from year to year and from season to season and cannot be relied on to pinpoint a position. Large lakes and rivers can be used as important aids to navigation and movement.

9-4. DISTANCE AND DIRECTION

When traveling from one point to another, it is necessary for you to know the distance between them beforehand, then plot your journey accordingly.

9-5. MEASURING DISTANCES

PACING

In temperate zones, pacing is the simplest method of measuring distance. In cold-weather areas, snow cover and the use of skis and snowshoes make pacing challenging, and it must be practiced repeatedly.

USE OF FIELD WIRE

The best method for accurate ground measurement is for two men to use a piece of salvage field wire 50 meters long. The lead man marches off in the desired direction trailing the wire behind him. The second man signals by jerking the wire when the first has gone 50 meters and then moves forward to the first man's position. The first man marks the spot and continues forward for another 50 meters. This is repeated until the desired distance is reached.

ESTIMATION

If no other method of measurement is available, the length of time spent in steady progress along a route may be used to estimate the distance traveled. The accuracy of this method depends on how well you know your marching speed over all types of terrain on foot, skis, or snowshoes.

9-6. DETERMINING DIRECTION

COMPASS

The use of the compass is the most common way to determine direction. The magnetic field becomes weaker as you get nearer to the magnetic poles and can be easily affected by other objects. It is important to make sure that the compass is not affected by iron fragments in the ground, vehicles, or metal objects on you.

THE SUN

There are two ways to determine direction by the sun:

- At noon, in the northern hemisphere, shadows fall to the north of objects; in the southern hemisphere, to the south of objects.
- Estimate direction by using a watch and the sun. In northern latitudes the hour hand is pointed toward the sun. An imaginary line halfway between the hour hand and 1200 will point south. In southern latitudes the procedure is different. Then point the 1200 toward the sun. An imaginary line halfway between the hour hand and 1200 will point north.

NATURE

Nature will help in determining direction as in the following samples:

- Single evergreen trees, growing away from other trees, will always be more bushy on the south side.
- The bark of birch and poplar trees is whitest on the south side and darkest on the north.
- You should know from which direction the prevailing winds blow. Trees will be bent in the direction the wind is blowing or expected to blow.

THE STARS

The North Star may be used to determine true north. This star can be easily found by its position relative to the Big Dipper. In latitudes under 70 degrees when traveling north, the North Star makes a good steering mark since its bearing is usually only 1 degree from true north and is never more than 2 1/2 degrees away. In higher latitudes above 70 degrees, the North Star is too high in the sky to indicate good direction.

A star near the horizon can be used with a bearing within 2 degrees of your compass course. However, it will only be good for up to 30 minutes when you are traveling north and only 15 minutes when traveling south. After these intervals, select another star.

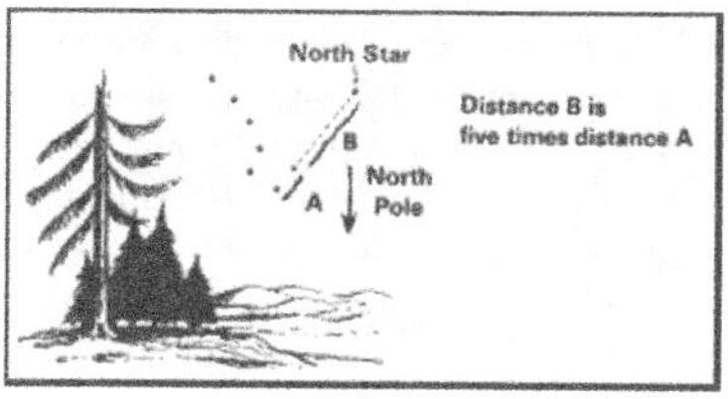

Figure 9-1. Determining direction using North Star.

South of the equator, the constellation Southern Cross will help you locate the general direction of south and any other direction. This group of four bright stars is shaped like a cross that is tilted to one side. The two stars forming the long axis, or stem, of the cross are called pointers. From the foot of the cross, extend the stem five times its length to an imaginary point. This point is in the general direction of south. From this point, look straight down to the horizon and select a steering point.

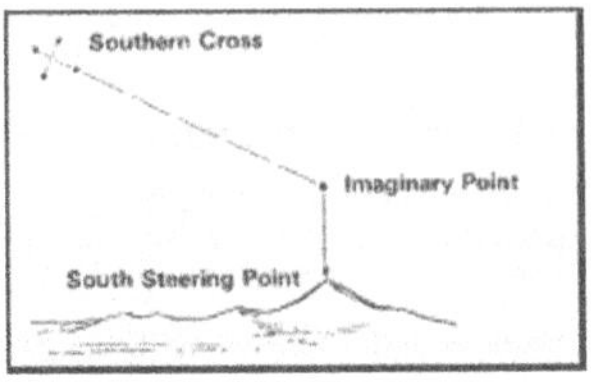

Figure 9-2. Determining direction using Southern Cross.

9-7. NAVIGATION BY DEAD RECKONING

Dead reckoning consists of plotting and recording a series of courses from a known point, each measured for distance and direction. These courses lead from the starting point to the destination. This lets the navigator determine his position at any time, either by following his plot or by comparing his actual position on the ground to his plotted course. All the details are previously estimated and recorded.

EQUIPMENT FOR DEAD RECKONING

- Maps are used for choosing the route and for plotting the route taken as the march progresses.
- Aerial photographs are usually used in addition to maps. They must be vertical photos taken during the same season in which they are to be used.
- The compass is used as the main direction-finding instrument.
- The protractor is necessary in dead reckoning for plotting direction and distance on the map, aerial photo, or chart. When using it, remember to convert magnetic azimuth to grid azimuth.
- A route card is used to outline the plan of the proposed march. Keeping a log is mandatory. It is used to record the distance and direction.
- The log is a written record of where the soldier has started, the distance and direction traveled, and his position at any time in relation to the planned route, destination, and starting point.

THE DRILL OF DEAD RECKONING

Know the starting point and destination and, having examined the map or aerial photo for the best route, prepare a route card. This describes each leg of the proposed route in terms of distance and direction. As the march continues, keep a careful record of each bearing taken and the distance, and record this in the log. There may be a change to the planned route because of terrain or enemy action. Therefore, the necessary changes are made using the map or aerial photo as a guide, and they are recorded in the log.

9-8. HINTS FOR THE NAVIGATOR

The navigator must keep the compass warm to speed up recording his position. When no steering marks exist to the front, he should march on a back azimuth. A steering mark may be a natural or man-made structure to the rear. If the steering mark is not clear, he must keep an eye on it after taking an azimuth so that it doesn't become lost. Never take magnetic azimuths in the area of metal objects.

When visibility is poor, only close-in steering marks will be available. Under these conditions try to pick up steering marks along the correct azimuth. This can be done correctly only when the route to each steering mark follows a straight line. Take compass checks often to ensure that the correct azimuth is being followed. Set the compass for a night course. A steady unshifting wind will also aid in keeping direction.

Depending on the area of the world, other signs can be used. For example, in the barren and semibarren lands of eastern Canada, snowdrifts point to the northwest and windblown trees to the southeast because of prevailing winds. In Scandinavia, scratches on rocks and cliffs caused by the glacial period (ice age) run northwest to southeast.

CHAPTER 10

Survival

"Each day the number of my dogs dwindled and the sleds were broken up to cook those animals that we ate ourselves."

Peary, Arctic Expedition, 1906

10-1. GENERAL

Your mission may take you to an arctic or subarctic region. You expect to have all personal equipment and other members of your unit with you wherever you go. But you could find yourself alone in a remote area with little or no personal gear.

10-2. ACTION WHEN LOST

Brief the soldier on the route before setting out on a march; he should memorize details if possible. A soldier can become separated from the main body of troops, especially if an action is fought. This applies mainly to night battle or bad weather. It can also happen if he is careless and march discipline is poor.

WHEN LOST WITHIN A KNOWN LOCALITY

If all is quiet and there are no battle sounds or aircraft to guide you to friendly lines, go back to the last known position. If you cannot go back, consider carefully the journey already made and the distances and directions from the starting point.

Opinions should be taken from the group. Search parties should be sent out to find the next known steering mark. Trails must be marked carefully so that the search party can return to guide the main group forward or to rejoin the group, should their search be unproductive. Meanwhile, the rest of the group should seek shelter.

If the search party fails to find any known location, perform the steps below, but remember to think clearly and keep calm.

CONDUCT WHEN LOST

When you are certain that you are lost:

- Stay together. Meet your emergency as a group; don't wander off as individuals; never allow stragglers.
- Take shelter. If you have a tent, set it up and get in it. If you do not have a tent, build one of the improvised shelters. The type of shelter is not important as long as you protect yourself from the weather.

- Keep warm. If you are lost in a forest, light a fire (depending on the enemy situation). If you are lost in an area where there are no trees, do not use your gasoline stove to heat your shelter. Use your stove only to prepare hot food and drinks. Conserve your fuel supply; you won't get any more. Get into your sleeping bag to stay warm.
- Check food supply. All the food and water in the group should be collected and rationed systematically. Supplement the food supply by shooting large animals and trapping or snaring small ones.
- Check communications. If you have a radio set, use it to contact your parent unit for help. If you are in a forest, prepare fires to attract search planes. Smoke is better by day and flame at night. Mirrors, ground signals, etc., are helpful in attracting search parties. Arrange necessary ground/air signals for search planes.
- Rest. The more you rest the less energy you expend and the less food you need.
- Adhere to survival tips. Carry safety matches inside the butt plate of your rifle or in your pockets. Carry candles, which are easy to light, provide heat and light, and help in starting a fire. If you have no matches, fire may be produced by striking the flint on your waterproof match box with the blade of a knife and collecting sparks in dry tinder. Dry moss and bark make good tinder to start your fire. Fires should be built in layers with sticks being laid out like the spokes of a wheel.
- Carry a two-cell flashlight. It can be seen from the air or on the ground as far as line-of-sight will permit.
- Ground/air signals can be made with small fires to form the identification characters (see Appendix C).
- Survival depends on clear thinking and resourcefulness. The main things to remember are to keep calm, think, try to help each other, keep together, and keep warm.

10-3. IMPROVISED SHELTERS

Sometimes tents or other regular shelters are not available. In summer, if the weather is mild, you may need protection only from insects. In the winter, however, you cannot stay in the open for long periods unless you are moving. In case of emergency, you must know how to protect yourself from the effects of the weather.

If natural shelters such as caves or rock ledges are available, they should be used. If they are not, a temporary improvised shelter must be built.

The type of shelter to be built depends on the equipment and materials available. By the proper use of materials available, some sort of shelter can be built during any season of the year. In open terrain a shelter can be built using ponchos, canvas, snow blocks, or other materials. Snow caves, snow trenches, snow houses, or snow holes may be built in the winter. In the woods, a lean-to is normally preferred to other types of shelter. In cold-weather areas, nature provides you with the means to prepare a shelter. Your comfort, however, greatly depends on your initiative and skill at improvising.

PONCHO SHELTERS

The simplest type of shelter can be made by pulling the poncho over your sleeping bag. For additional comfort, various types of shelters and lean-tos can be made by attaching your poncho to trees, tree branches, or poles.

To build a one-man shelter from one poncho, spread the poncho on the ground, hood side up. The hood opening must be tightly closed by adjusting and tying the drawstrings. The poncho is raised in the middle of its short dimension to form a ridge, with the stake out at the corners and sides. Snow, sod, or branches are used to seal the side and one end of the shelter to provide additional protection from the wind and to retain heat.

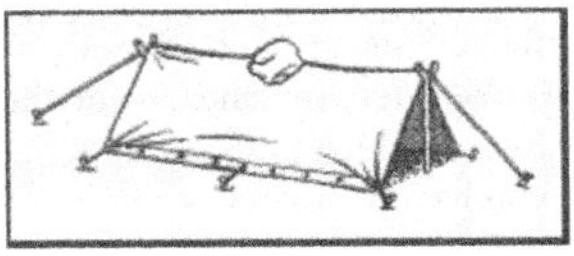

Figure 10-1. One-man shelter.

A two man shelter can be made from two ponchos. Spread the ponchos on the ground with the hood side up and the long sides together so that the snap fastener studs of one poncho may be snapped into those of the other poncho, The ponchos are raised where they are joined to form a ridge. The shelter is then staked out at the sides and corners. A third poncho may be snapped into the other ponchos to form a ground cloth.

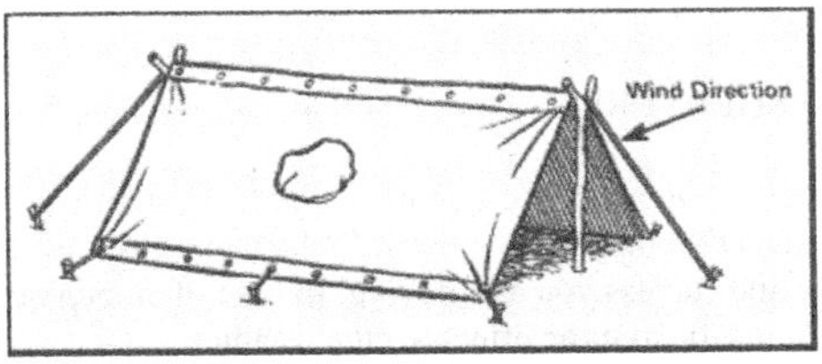

Figure 10-2. Two-man shelter.

LEAN-TO SHELTER

The lean-to shelter is built in forested areas and is made of trees and tree limbs. A poncho, a piece of canvas, or a parachute, in addition to boughs, may be used for covering. Depending on the number of people to be sheltered, two types of lean-tos (single and double) are built. Boughs should be laid in the shelter like overlapping shingles (the same as on a live tree).

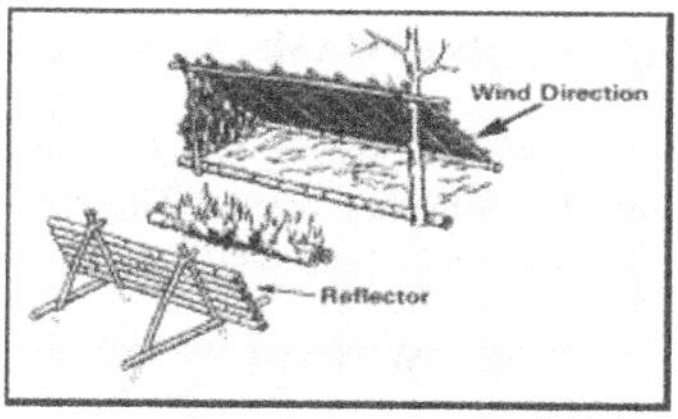

Figure 10-3. Single lean-to.

To save time and energy, use two trees the right distance apart and sturdy enough to support the crosspiece. If suitable trees are not available, use two forked poles or build two A-frames to hold the crosspiece. A large log is laid at the back for added height. Stringers about 3 meters long and 10 centimeters in diameter are then placed 45 centimeters apart from the crosspiece over the top of the log in the rear. Available coverings, ponchos, boughs, and so forth are then placed on top of the stringers. A double lean-to is made by building two single lean-tos facing each other with the fire in between.

Any kind of an open fire can be used to heat the lean-to. The best kind for a large lean-to is the log fire. Precautions must be taken to prevent the fire from burning too hot and burning down the shelter. A fire reflector may be built of green logs or poles to reflect the heat into the shelter and to serve as a windbreak to keep the fire burning steadily.

TREE SHELTERS

In wooded areas, the tree-pit shelter furnishes temporary protection. Select a large tree with thick lower branches and surrounded with deep snow. Enlarge the natural pit around the tree. The walls and floor are lined with branches and boughs. Try not to disturb the snow that may be on the branches.

An emergency shelter for one man can be built by cutting down a coniferous tree at a point about 1 meter from the ground.

Figure 10-4. Tree-pit shelter.

After the tree has fallen, the branches on the underside are trimmed away and the cut material used for insulation on the ground.

WIGWAM

A wigwam or tepee can be built in wooded areas by tying a number of poles near the top and spreading them at the bottom to form a large circle. This framework is then covered with ponchos, a parachute, or any other suitable material.

SNOW WALL

In open terrain with snow and ice, a snow wall may be built for protection from strong winds. Blocks of compact snow and ice are used to build a windbreak. Be careful not to get covered up by drifting snow, which may accumulate on the lee side of the wall.

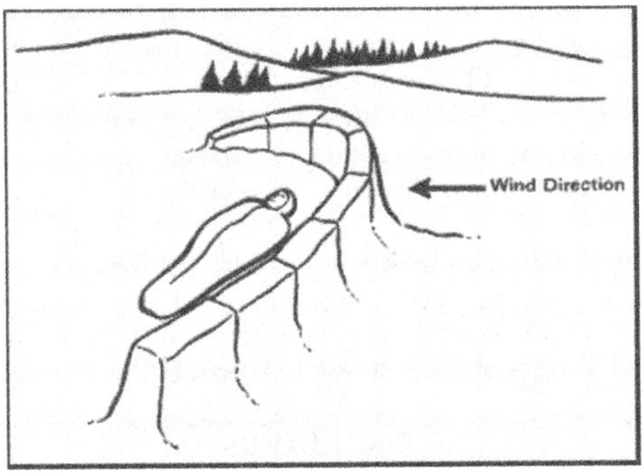

Figure 10-5. Snow wall.

SNOW TRENCH

A snow trench provides shelter quickly. It is built by burrowing into a snowdrift or by digging a trench in the snow and making a roof of ponchos or ice and snow blocks supported by skis, ski poles, or snowshoes. Boughs can be used for covering the roof and for the bed. Figure 10-6 shows one method of roofing over the trench by leaning snow blocks against each other.

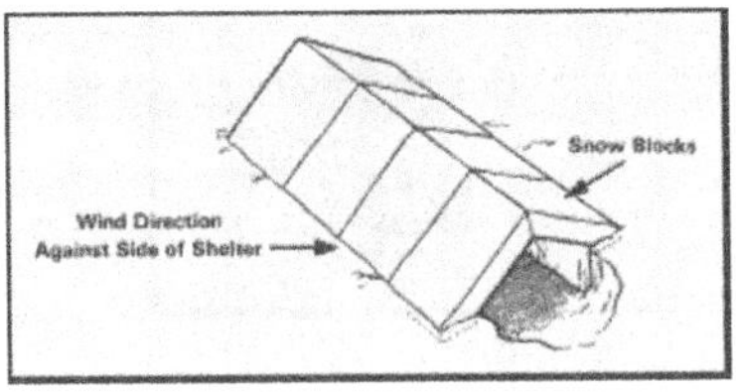

Figure 10-6. Snow trench.

SNOW CAVES

A snow cave can be used as an improvised shelter in the open areas where deep and compacted snow is available. Normally, a suitable site can be located on the lee side of a steep ridge or riverbank where drifted snow collects, and it is wind packed.

The tunnel entrance must give access to the lowest level of the chamber. This is the bottom of the pit, where cooking is done and equipment is stored. The snow cave must be high enough to provide comfortable sitting space. The sleeping area must be on a higher level than the highest point of the tunnel entrance, where there is warmer air.

The roof must be arched for strength so that drops of water forming on the inside will not fall on the floor. This water will follow along the curvature of the roof and sides, glazing over the walls when frozen. The roof must be at least 30 centimeters thick. The size of the snow cave depends upon the number of men that will occupy it. A large cave is usually warmer and more practical to build and maintain than several small caves.

The shape of the snow cave can be varied to suit conditions. When the main cave is built, short side tunnels are dug to make one- or two-man sleeping rooms, storage space, latrine, or kitchen space.

In building the snow cave, locate a deep snowdrift at least 3 meters deep. Newly fallen, powdery, or loose snow should not be used. The depth of a snowdrift may be tested with a sharpened branch or young tree about 4 meters long, or with the ski or shorter ski pole. The entrance must be chosen carefully so that wind will not blow into the cave or block the entrance by drifting snow. The entrance should be built so that it is about 45 degrees from the downwind side. A small tunnel is burrowed directly into the side of the snowdrift for 1 meter. A chamber is excavated from this tunnel. Excavation is done to the right and left, so that the length of the chamber is at right angles to the tunnel entrance. Personnel doing the digging will become wet from perspiration and from the snow inside the cave. They should wear the minimum amount of clothing to ensure that they have a change of dry clothing when finished.

The cave can be heated with the one-burner gasoline stove or with candles. The fires should be put out when personnel inside the cave are sleeping, thus reducing the danger of fire, asphyxiation, or carbon monoxide poisoning. If the weather is severe and it is necessary to have a fire while sleeping, an alert fire guard must be posted in each cave. The ventilation holes must be inspected every 2 or 3 hours to ensure that they have not become clogged by snow or ice.

To ensure that the cave is warm, the entrance should be blocked with a rucksack, poncho, or snow block when not in use. All available material such as ponchos, cardboard from ration boxes, brush, and boughs should be used for ground insulation.

Walking on the roof may cause it to collapse. At least two ventilator holes are used, one in the roof and one in the door. A ski pole can be stuck through the roof ventilator to clear it from the inside. Extra care must be taken to keep air in the cave fresh when cooking or heating. The entrance should be marked by placing a pair of skis, or other equipment, upright on each side of the entry way.

Building a snow cave requires practice. Large shelters are usually more efficient for squad-size units and larger units. The most critical decision is when and where to stop and build the shelter.

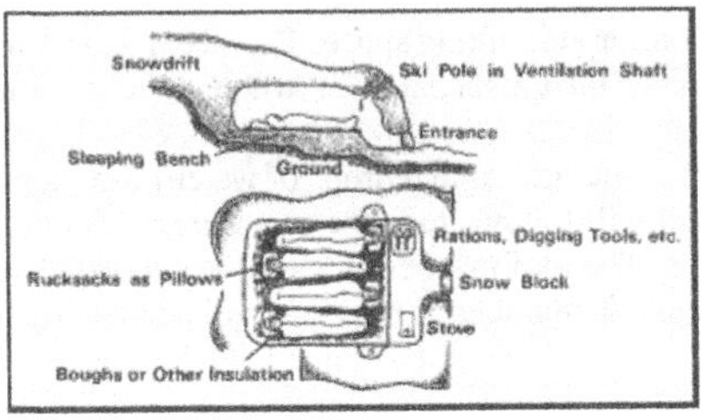

Figure 10-7. Snow cave.

INDIVIDUAL SNOW HOUSE

If you are alone and have need of a shelter quickly, you can build a small individual snow house. It is box-shaped, and the only building materials needed are snow blocks. Cut snow blocks 40 by 40 by 80 centimeters. Lay them on edge to a length of 2.5 meters and a width of 1 meter. Build the walls up to a height of about 1 meter. When you have the walls built, make an A-shaped roof by leaning together pairs of blocks. Fill in the ends with snow blocks and chink all the crevices with soft snow. Cut a hole in the end of the leeward side large enough to crawl into. When you are inside, make one or two small holes near the top for ventilation. Block up the end with a snow block or with your rucksack.

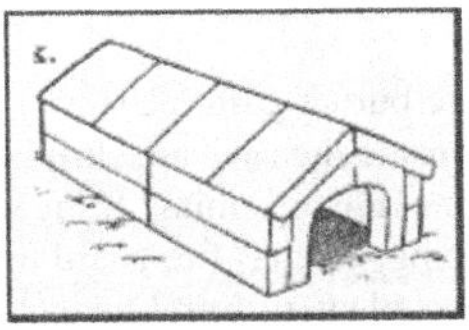

Figure 10-8. Individual snow house.

SNOW HOUSE OR IGLOO

An igloo makes an excellent shelter in snow and glacier areas when other materials are not available or when the snow is so hard that it can be cut into blocks. Cracks in the igloo are then covered with loose snow. To increase the head room and to prevent water droplets from forming inside the igloo, the corners of the blocks may be cut off to form a round and smooth interior.

A shelter may be built over the entrance in the same manner as the igloo, and it should be 1 to 2 meters long and about 1 meter wide and 1 meter high. This shelter provides more protection and warmth inside the igloo and a place to store equipment. Blocks of snow or other material may be placed at both the outside and the inside entrances of the tunnel for more warmth. Make a ventilation hole near the top and at the same height as the fifth or sixth row of snow blocks. The hole can be kept open by placing a ski pole or long stick into it and clearing it out as needed. The igloo is heated with a one-burner stove or with candles.

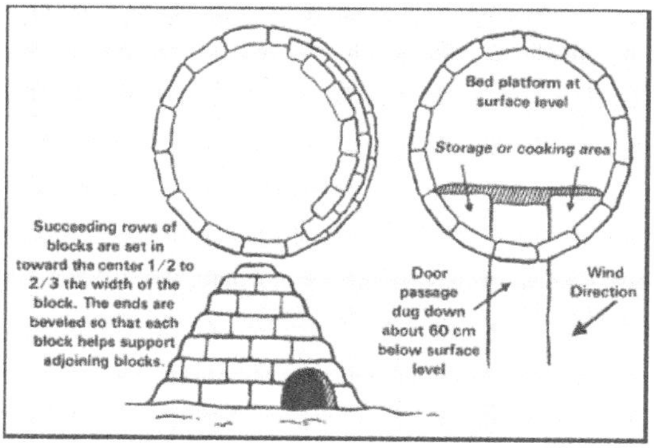

Figure 10-9. Snow house.

10-4. FOOD

The chances of finding different types of food vary in cold-weather areas depending on the time of year and the place. Shores normally are scraped clean of all animals and plants by winter ice. Food can still be found north of the timberline, even when such foods as mice, fish, and grubs are not available.

STORAGE AND PRESERVATION

If a large animal is killed or a large number of small game are found, you should store or preserve some of the meat for future use. Freezing is the best way to preserve fresh meat or fish. It may be necessary to protect your supplies from small animals. This can be done by hanging these supplies about 2 meters from the ground.

FISH

There are few poisonous fish in cold-weather areas, but some fish, like the sculpin, lay poisonous eggs. The black mussel may be poisonous at any season and is as deadly as strychnine. In cold regions do not eat shark meat; it is poisonous. In coastal streams and rivers, salmon moving upstream to spawn may be plentiful; however, their flesh deteriorates as they travel away from the coast, making them unfit to eat except in an emergency. A salmon that has been in fresh water long enough to turn a brilliant red is decaying and may be unfit to eat.

Coastal waters are rich in all seafood. Grayling, trout, white fish, burbot, and pike are common to the lakes, ponds, and arctic coastal plains of North America and Asia. River snails or freshwater periwinkles are plentiful in the rivers, streams, and lakes.

Fish can be speared, shot, netted, hooked, caught by hand, or stunned by a rock or club. Fish hooks can be improvised from pins, pieces of wire, or any other metal object that can be bent into a hook. Pieces of meat, insects, or minnows can be used for bait. Some fish will nibble at any small object that hits the water. Cod will swim up to investigate strips of cloth or bits of metal or bone, and they are easily caught. A net can be made out of stout twine or from the inner strands of parachute suspension lines. These strands can also serve as fishing lines.

LAND ANIMALS

Deer, caribou, reindeer, musk-oxen, moose, elks, mountain sheep, goats, bears, and other such animals are found in most cold-weather areas.

Tundra animals include rabbits, mice, lemmings, ground squirrels, wolves, and foxes. Where there are trees, the porcupine is often encountered. The porcupine can be easily clubbed or shaken out of trees. Pick up a porcupine only after it is dead.

Hunting is better during the early morning or late evening when the animals are moving to and from feeding and bedding grounds and water. Large game such as moose and caribou should be shot just back of the front shoulder or, if the range is short and you are sure of your shot, in the neck. These large animals are fairly easy to stalk and kill, and their skins are also useful for survival. Bears should be shot just forward of the front shoulder to stop them or, if you are sure of your shot, in the neck. To successfully hunt these land animals you should know something of their characteristics:

- Caribou or reindeer are usually very curious. It is possible to attract them near enough for a shot by waving a cloth and moving slowly toward them on hands and knees.
- The technique of imitating a four-legged animal may also cause a wolf to come closer to a hunter.
- Moose may be found in heavy brush or around the shoreline of lakes.
- Mountain goats and sheep are cautious and hard to approach. They can be surprised, however, by moving quietly downwind while they are feeding. If possible, stay on higher ground than they are on.

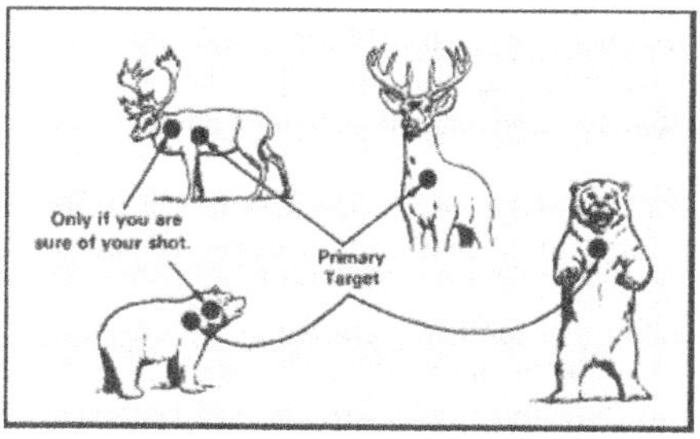

Figure 10-10. Hunting land animals.

- Musk oxen leave cattle-like tracks and droppings. When alarmed, they group together in a circle with their heads facing out and remain in that position unless approached, then one or more bulls may charge.
- Bears are always dangerous. A wounded bear is most dangerous and should not be followed into cover. The polar bear is a tireless, clever hunter with good sight and an extraordinary sense of smell. Be careful that you don't become the hunted instead of the hunter.
- Rabbits often run in circles and return to the same place when they are frightened. A running rabbit can sometimes be stopped by whistling. Snares made of wire and placed on small game trails, arranged so that the animal will place his head through the loop, are an excellent means of catching rabbits.
- The quick-kill method can be used on most small animals and sometimes on larger animals after they have been shot and are on the ground. The method is simple and consists of cutting the animal's throat with a sharp knife. This system not only kills the animal but bleeds him at the same time.

Figure 10-11. Snaring a rabbit.

SEA ANIMALS

During winter and spring, sea mammals — seals, walruses, and polar bears — are found on the frozen pack ice and on floes in open water.

Seals are hard to approach but can be stalked. Keep downwind and avoid sudden moves. A white camouflage suit will help. Advance only when the animal is sleeping. Do not eat the liver of the bearded seal, because its high vitamin A content may cause you to be sick.

Walruses are found on pack ice and must be approached by boat. Because of its large size and boldness, the walrus is one of the most dangerous animals in the arctic.

Polar bears are found in almost all arctic regions but rarely appear on land. Avoid them if possible. If it is necessary to kill one for food, do not eat the liver. The liver is high in vitamin A, which may cause you to be sick. Never eat any bear meat unless it is cooked; it is always diseased.

BIRDS

The breeding ground of many birds is in the arctic. Ducks and geese build their nests near ponds on the coastal plains during the summer and provide an important source of food. Grouse and ptarmigan live in mountainous terrain and brush-covered areas in arctic and subarctic areas. Sea birds may be found on cliffs or small islands on the coast. Sea birds, ravens, and owls are useful for food.

In winter, owls, ravens, grouse, and ptarmigans are the only birds available. Ptarmigans and grouse are easily approached and are very tame. Although they are hard to locate because of the protective coloring, they provide a good source of food because they can be killed with a stone or a club.

PLANTFOOD

Most plants that grow in cold regions can be eaten. The water hemlock and the baneberry are the only poisonous plants. The water hemlock is one of the world's most poisonous plants. It can be recognized by where it grows, always in wet ground, and by the following characteristics: a hollow, partitioned bulb at the base of a hollow stem, spindle-shaped roots, and a foul odor, which is very noticeable in the root and bulb. The water hemlock is abundant in marshes near southern beaches and around marshy lakes in interior river valleys. It is never found on hillsides or dry ground. The baneberry is normally found on hillsides or in forested bog areas. Some mushrooms are poisonous and should not be eaten. Although the cold-region areas have many kinds of berries that are edible, most will not be available in the winter.

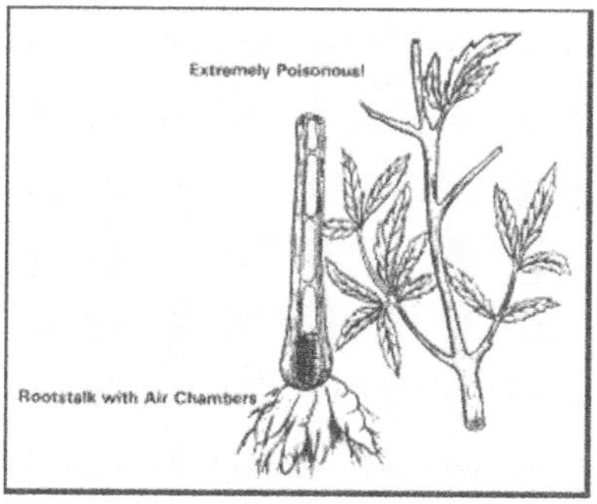

Figure 10-12. Water hemlock.

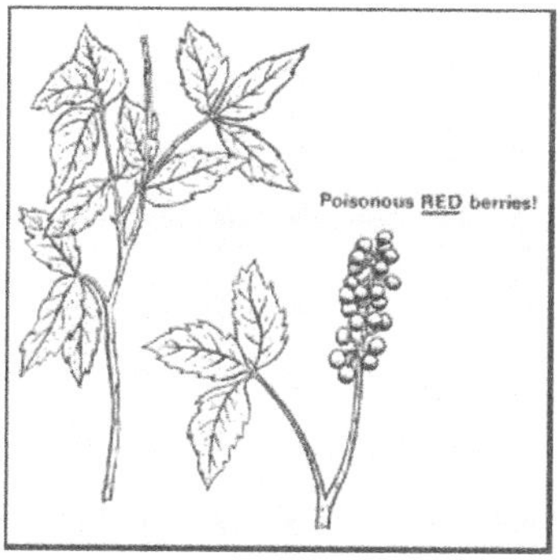

Figure 10-13. Baneberry.

10-5. USEFUL KNOTS

A rope is only a limp coil of hemp or nylon. It is useful only when you are able to attach it to itself or another object. The six basic knots described below will fit all your needs for tying rope together or onto equipment or personnel.

THE HALF HITCH

The half hitch is used to secure the end of a rope to some other object or to secure other knots.

Figure 10-14. Half hitch.

THE OVERHAND KNOT

The overhand knot is used to make a knotted rope for a handline or to temporarily whip the end of a rope to keep it from unraveling.

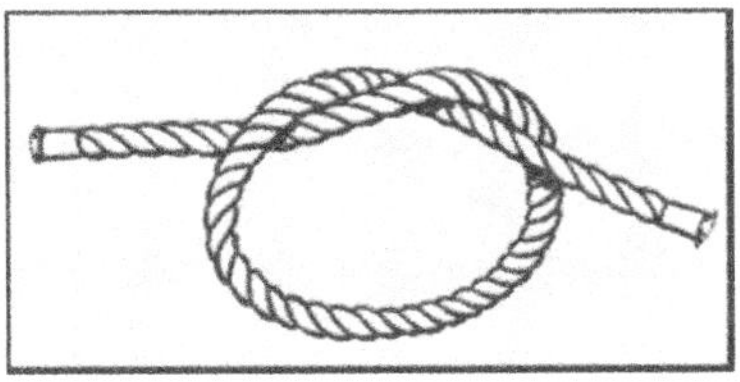

Figure 10-15. Overhand.

THE SQUARE KNOT

The square knot is used to tie the ends of two ropes of equal width together. The knot should be secured by tying a half hitch on each side of the knot. If properly tied, the square knot will not slip or jam. Be careful to follow Figure 10-16 or you may end up with a "granny knot," which can slip and jam.

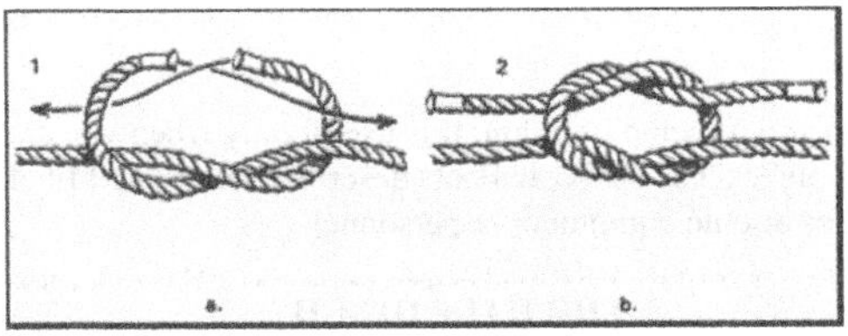

Figure 10-16. Square.

THE CLOVE HITCH

The clove hitch is the most useful of the hitches. It can be used to tie the end of a rope to a post or eye, or it can be used to secure the middle of a rope without using the ends.

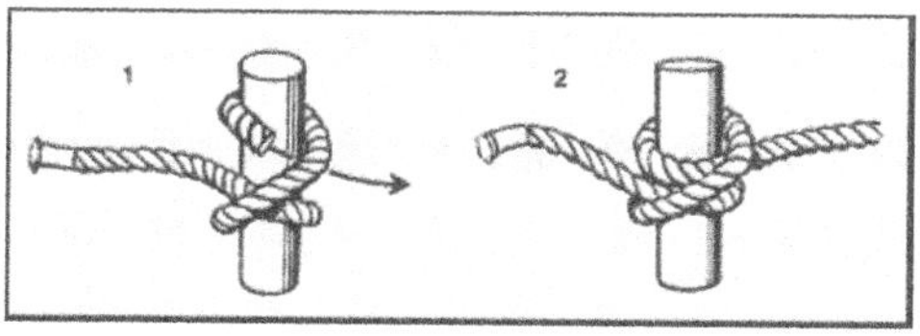

Figure 10-17. Clove hitch.

THE DOUBLE SHEET BEND

The double sheet bend is used for tying two ropes together of different width.

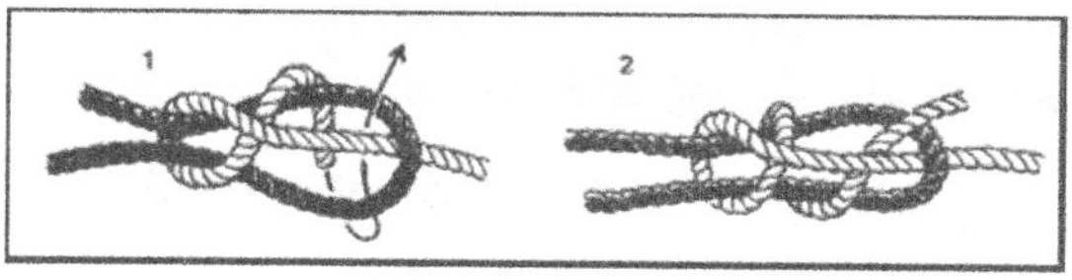

Figure 10-18. Double sheet bend.

THE BOWLINE

The bowline is used to form a loop on the end of the rope that will not slip.

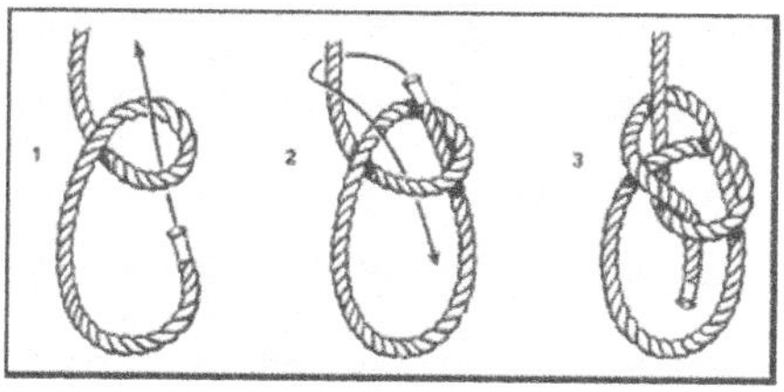

Figure 10-19. Bowline.

Remember the qualities of a good knot: it is easy to tie, it does not jam and become hard to untie, and it does not slip when weight is put on it.

APPENDIX B

Approximate Measurement Conversion

Factors

Multiply	*By*	*To Obtain*
Centimeters	.03	Feet
Centimeters	.4	Inches
Meters	3.3	Feet
Meters	40.0	Inches
Meters	1.1	Yards
Kilometers	.6	Miles
Knots	1.2	MPH
Inches	2.5	Centimeters
Feet	30.0	Centimeters
Yards	.9	Meters
Miles	1.6	Kilometers
MPH	.9	Knots

For ease in conversion the above factors have been rounded off.

References

Suggested Readings

DA Pam 20-201	Military Improvisations During the Russian Campaign
DA Pam 20-203	Russian Combat Methods in World War II
DA Pam 20-231	Combat in Russian Forests and Swamps
DA Pam 20-234	Operations of Encircled Forces: German Experiences in Russia
DA Pam 20-236	Night Combat
DA Pam 20-240	Rear Area Security in Russia: The Soviet Second Front Behind the German Lines
DA Pam 20-269	Small Unit Actions During the German Campaign in Russia
DA Pam 20-271	The German Northern Theater of Operations, 1940-1945
DA Pam 20-290	Terrain Factors in the Russian Campaign
DA Pam 20-292	Warfare in the Far North
FM 9-207	Operation and Maintenance of Ordnance Materiel in Cold Weather (0° to -65°F)
FM 31-70	Basic Cold Weather Manual
FM 31-71	Northern Operations
FM 90-6	Mountain Operations
FM 90-11 (TBP)	Winter Operations
TC 90-11-1	Military Skiing

"Perhaps the most valuable result of all education is the ability to make yourself do the things you have to do when it ought to be done, whether you like it or not."

Hosley

Glossary

ALICE	all-purpose lightweight individual carrying equipment
C	Celsius
CLP	cleaner lubricant preservative
ECWCS	extended cold-weather clothing system
F	Fahrenheit
FM	field manual
GP	general purpose
SFP	synthetic fiber pile
SOP	standing operating procedure
TBP	to be published
VB	vapor barrier